OR at wORk
Practical experiences of operational research

OR at wORk
Practical experiences of operational research

EDITED BY

LEONARD FORTUIN,
PAUL VAN BEEK

AND

LUK VAN WASSENHOVE

Taylor & Francis
Publishers since 1798

UK Taylor & Francis Ltd, 1 Gunpowder Square, London, EC4A 3DE
USA Taylor & Francis Inc., 1900 Frost Road, Suite 101, Bristol, PA 19007

British Library Cataloguing in Publication Data
A catalogue record for this book is available from the British Library.
ISBN 0 7484 0455 4 (cased)
ISBN 0 7484 0456 2 (paperback)

Library of Congress Cataloguing in Publication data are available

Cover design by Amanda Barragry
Typeset in Times 10/12pt by Keyset Composition, Colchester, Essex
Printed in Great Britain by T.J. Press (Padstow) Ltd

Contents

vi   CONTENTS

Preface

Approximately 10 years ago, three colleagues found, almost by accident, that they shared a common interest in promoting successful operational research (OR) practice.[1] They decided to team up. The 'OR-*troika*', as they came to be known later, was born. Over the years, the *troika* developed a number of initiatives, none of them earth shattering since they could be no more than a hobby, gobbling up personal free time of otherwise busy professionals. The *troika*, among other things,

- wrote articles on the practice of OR to be published in general management journals, professional (e.g. engineering) journals and in the popular press (newspapers etc.);
- organized sessions on OR practice and the process of conducting OR studies at conferences (EURO,[2] INFORMS,[3] IFORS[4]);
- initiated the 'OR Practitioners Network', a global network of OR professionals (which in 1994 became a working group within EURO);
- initiated a competition for 'The Best Applied Paper'. The first competition was run concurrently with the EURO XIV Conference in Jerusalem (July 1995).

All these initiatives were aimed at making OR more visible and at promoting good applied work. Ten years is a long time. Many initiatives were successful and have been institutionalized. New, younger and more dynamic colleagues have taken over our efforts. The *troika*'s (mini)crusade is over. However, before disbanding officially, the *troika* set itself a final goal: to produce a *pièce de résistance* in the form of a book filled with successful OR applications and a few reflective chapters about how our beloved discipline should be practised. It took a while for the child to be born but here it is at last. You are looking at it.

Leonard Fortuin
Paul van Beek
Luk Van Wassenhove

Notes

[1] The need to promote successful OR practice through professional societies and education is obviously perceived differently in different countries or regions. For example, the UK has an active and successful practitioners' community as well as a society with a rich tradition in and attention for the process of OR.

[2] Association of European Operational Research Societies.

[3] Institute for Operations Research and the Management Sciences.

[4] International Federation of OR Societies.

Contributors

P. van Beek
Agricultural University Wageningen, Dept of Mathematics, Section OR, Dreijenlaan 4, NL 6703 HA Wageningen, The Netherlands

J. F. Benders
Engelsbergenstraat 28, 5616 JC Eindhoven, The Netherlands

C. G. E. Boender
ORTEC Consultants bv, Groningenweg 6-33, 2803 PV Gouda, The Netherlands and Erasmus University Rotterdam, PO Box 1738, 3000 DR Rotterdam, The Netherlands

G. D. H. Claassen
Agricultural University Wageningen, Dept. of Mathematics, Dreijenlaan 4, NL 6703 HA Wageningen, The Netherlands

Charles J. Corbett
Owen Graduate School of Management, Vanderbilt University, Nashville, TN 37203, USA

Joachim Daduna
Fachhochschule Konstanz, Fachbereich Wirtschafts- und Sozialwissenschaften, Brauneggerstrasse 55, D-78462 Konstanz, Germany and Badensche Strasse 49, D-10715 Berlin, Germany

R. Dekker
Econometric Institute, Erasmus University Rotterdam, PO Box 1738, 3000 DR Rotterdam, The Netherlands

M. C. Dijkstra
Bouwcenter Groep, PO Box 296, NL-3440 AG Woerden, The Netherlands

H. A. Fleuren
Centre for Quantitative Methods (CQM bv), PO Box 414, NL 5600 AK Eindhoven, The Netherlands

L. Fortuin
Eindhoven University of Technology, PO Box 513, Paviljoen F.18, NL 5600 MB Eindhoven, The Netherlands

Y. van der Graaf
Utrecht University and Academic Hospital, Department of Clinical Epidemiology, Huispost L00.315, PO Box 85500, 3508 GA Utrecht, The Netherlands

J. Dik F. Habbema
Erasmus University Rotterdam, Department of Public Health, PO Box 1738, 3000 DR Rotterdam, The Netherlands

E. M. T. Hendrix
Agricultural University Wageningen, Department of Mathematics, Dreijenlaan 4, 6703 AH Wageningen, The Netherlands

D. den Hertog
Centre for Quantitative Methods (CQM bv), PO Box 414, NL 5600 AK Eindhoven, The Netherlands

L. A. van Herwerden
Thoraxcenter Dijkzigt Hospital BD156, Erasmus University Rotterdam, Dr. Molenwaterplein 40, 3015 GD Rotterdam, The Netherlands

L. Hordijk
Centre for Environment and Climate Studies, Wageningen Agricultural University, PO Box 9101, NL-6700 HB Wageningen, The Netherlands

C. A. J. Hurkens
Dept. of Mathematics and Computing Science, Eindhoven University of Technology, PO Box 513, HG 9.29, NL 5600 MB Eindhoven, The Netherlands

Karl Inderfurth
Otto-von-Guericke-Universität Magdeburg, Fakultät für Wirtschaftswissenschaft, Lehrstuhl für Produktion und Logistik, Universitätsplatz 2, D-39106 Magdeburg, Germany

L. G. Kroon
Rotterdam School of Management, Erasmus University Rotterdam, PO Box 1738, NL 3000 DR Rotterdam, The Netherlands

C. M. H. Kuijpers
Dept. of Computer Science and Artificial Intelligence, University of the Basque Country, PO Box 649, 20080 San Sebastian, Spain

R. W. J. Lukkassen
Centre for Quantitative Methods (CQM bv), PO Box 414, NL 5600 AK Eindhoven, The Netherlands

Dirk Meier-Barthold
Otto-von-Guericke-Universität Magdeburg, Fakultät für Wirtschaftswissenschaft, Lehrstuhl für Produktion und Logistik, Universitätsplatz 2, D-39106 Magdeburg, Germany

J. B. M. Melissen
Philips Research Laboratories, Prof. Holstlaan 4, NL 5656 AA Eindhoven, The Netherlands

J. H. P. van der Meulen
Dept of Clinical Epidemiology and Biostatistics, Academic Medical Centre – University of Amsterdam, Meibergdreef 9, 1105 AZ Amsterdam ZO, The Netherlands

J. A. E. E. van Nunen
Rotterdam School of Management, Erasmus University Rotterdam, PO Box 1738, NL 3000 DR Rotterdam, The Netherlands

W. J. A. M. Overmeer
International Business Department, Stern School of Business, New York University, 44 West 4th Street, New York 10012, USA and INSEAD, Boulevard de Constance, F 77305 Fontainebleau, France

E. H. Poot
De Verenigde Koelhuizen Hobabo bv, Hyacinthenlaan 16, NL-2182 DE Hillegom, The Netherlands

C. F. H. van Rijn
Beeckzanglaan 1c, 1942 LS Beverwijk, The Netherlands

J. Roosma
R. F. Burtonlaan 2, NL 2803 EW Gouda, The Netherlands

M. Salomon
Erasmus University Rotterdam, PO Box 1738, 3000 DR Rotterdam, The Netherlands

G. Schepens
Facultés Universitaires Notre-Dame de la Paix, 8 Rempart de la Vierge, B-5000 Namur, Belgium and Beyers Innovative Software, Michielssendreef 42, B 2930 Brasschaat, Belgium

L. Schepens
Beyers Innovative Software NV, Michielssendreef 42, B 2930 Brasschaat, Belgium

E. W. Steyerberg
Department of Public Health, Erasmus University, PO Box 1738, 3000 DR Rotterdam, The Netherlands

A. Van Looveren
Beyers Innovative Software NV, Michielssendreef 42, B 2930 Brasschaat, Belgium

L. N. Van Wassenhove
INSEAD, Boulevard de Constance, F 77305 Fontainebleau, France

Manfred Völker
HanseCom GmbH, Bereich MS, Spohrstrasse 6, D-22083 Hamburg, Germany

W. H. M. Zijm
University of Twente, Department of Mechanical Engineering, PO Box 217, NL 7500 AE Enschede, The Netherlands

CHAPTER ONE

Introduction

L. FORTUIN, P. VAN BEEK and L. N. VAN WASSENHOVE

1.1 Why this book?

Operational Research (OR) is an applied discipline. It is concerned with real-life problems and with people who need to deal with these problems. It is, of course, just one of many approaches to real-life problem solving characterized by an analytical (i.e. model- and data-based) way of looking at things.[1] It is precisely this analytical side of OR that can create problems. Indeed, too much emphasis on mathematical tools and techniques may severely hamper the usefulness of OR to human decision makers working with real people in real organizations. Good practice in our discipline means appropriate analytical skills as well as adequate softer skills (teamwork, facilitation, etc.).

Learning all the mathematical tools and how to apply them is one thing but learning how to conduct a successful OR project is quite another. This is partly due to the fact that OR teachers are often tools- and technique-oriented people without an industrial background. They have little practical experience and are uncomfortable with talking about the softer process skills involved in managing real-world projects. Moreover, their students are also quite often science students who prefer to stick to the mathematics and not be bothered by the 'touchy–feely' stuff. Obviously, business school students often come from non-quantitative backgrounds and are put off extremely quickly by a purely analytical approach to OR teaching. Needless to say this can seriously damage the perception of our discipline and its usefulness to business and society!

Teaching OR practice can be perceived as an oxymoron. One essentially becomes a good OR practitioner by doing OR projects and building experience. Nothing beats practice if one wants to be(come) a successful practitioner! However, this does not preclude other ways of learning about

OR practice. Good case descriptions, i.e. stories about real projects, can be extremely useful. They can help prepare students and novices for a successful start to their careers and they can also be useful to seasoned practitioners in benchmarking themselves against each other and in exchanging best practice.

Good and rich descriptions of successful practice can give students as well as potential users (e.g. managers) a much better picture of what OR is all about, how OR projects are typically conducted, what is expected of project team members and what can be reasonably expected from OR as one (of the many) way(s) of approaching a problem. Cases can and should give (future) practitioners and users of OR a view of the richness of the discipline and its variety. The latter is definitely a major objective of this book.

Unfortunately, a lot of what the OR community calls cases are nowhere near the definition of a case as a rich description of a successful project. OR cases are mostly unlike typical business school cases and are rather linear and dry descriptions of how a technique was applied to a particular environment. Their emphasis is often on the peculiarities of the techniques, rather than on the actual decision environment. Typically, the problem is well defined and given, and project management issues, in particular the softer ones, are downplayed or even completely ignored.

Consequently, OR teaching, OR textbooks and OR casebooks bear the risk of conveying simplistic and potentially damaging images of the discipline. One such image is that OR practice is about the application of tools and techniques to well-defined and given problems. It is much more than that. Another is the image of OR as a linear set of steps (problem description, model building, data gathering, selecting a solution method, validating model and solution method, sensitivity analysis and implementation of the solution; see Figure 1.1). Obviously, the latter image is much too sterile. OR practice is also about fuzzy dynamic situations with people having different goals, where the project, the people and the goals may change as the project evolves.

Summarizing, cases are an essential component in transferring OR practice. They can be of help in learning the tricks of the trade and/or in appreciating the richness, variety, capability and limitations of the discipline. Hence casebooks are essential. However, many OR cases are too narrow in scope and may therefore convey incomplete and perhaps misleading images of what it takes to be a good OR practitioner. There is a need for cases that downplay the techniques and how they were applied but emphasize the process of an OR intervention and how it can support change in an organization.

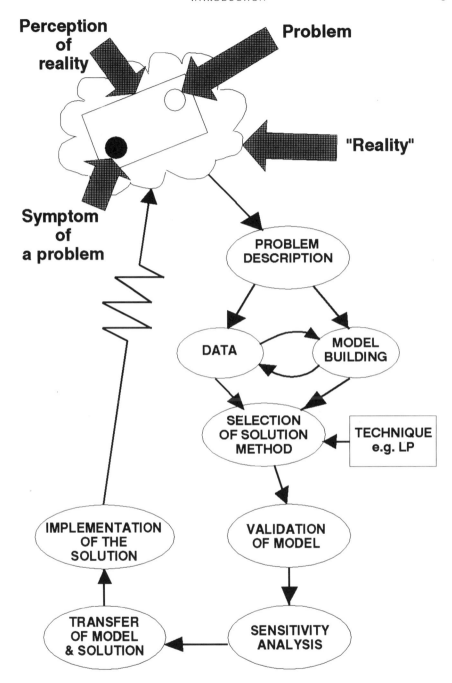

Figure 1.1 Conventional 'linear view' of an OR project (see Fortuin *et al.*, 1992).

1.2 **Another, but a different, book**

We started this book with the above in mind. It would have to contain cases that are somehow different from the standard format. Our instruction to authors read:

> The case studies should be interesting and pleasant to read. Emphasis should be on the process of problem solving and its results, rather than on techniques. Special attention should be paid to readability: the style should be fresh, avoiding jargon or details that can only thrill the initiated few.

We went as far as to suggest to authors a detailed outline on how to structure their case descriptions (Table 1.1). Some authors picked up the challenge and did a great job. Others stayed closer to traditional OR case descriptions. Old habits die hard.

We realized that even richer and more process-oriented case descriptions had their inherent limitations in conveying OR practice. So we decided to hit the road and interview successful OR practitioners about their practice as well as their clients about how they perceived the intervention of the OR practitioner. We were pleasantly surprised by the willingness of the practitioners to talk about the tricks of their trade and by the many useful insights we were able to gather from listening to practitioners reflect upon what it is they actually do (as opposed to what they write down in a case description!). The results of this exploratory investigation are described in Chapter 18. That chapter nicely complements the already richer and non-typical case descriptions in this book. Together they attempt to make the book different from most casebooks on OR by allowing the reader to capture much more the important process issues involved in the practice of OR.

Another element that sets this book apart from the usual casebook is its richness and variety of applications showing the power and usefulness of the discipline. Many casebooks emphasize resource allocation problems with a special focus on linear programming as a basic methodology. We provide a *clustering* of OR applications in the next section. Apart from being useful for didactic purposes, our application clusters allow us to de-emphasize the classical OR applications found in many casebooks and to present a richness of cases in other areas. This variety of successful applications in very different fields testifies to the power and health of OR as a discipline.

We sincerely hope that the richness and diversity of the applications will show that there is ample room for very different types of OR workers. We should be proud of the variety of our discipline and cultivate it. There is a lot of room for cross-fertilization. This would be much more beneficial to the discipline and its workers than many of the current debates and turf wars. We quietly hope that our book will contribute modestly to more tolerance, understanding and, who knows, perhaps cooperation between different types of OR workers active in the different application clusters to be described in the next section.

Table 1.1 Instructions to authors

The case studies to be described should be interesting for the proposed audience and pleasant to read. Emphasis should be on the *process* of problem solving and its *results*, rather than on *techniques*. Special attention is required for the readability: the style should be fresh, avoiding jargon and other elements that only can thrill the insiders. There is ample space for the aesthetic and pedagogical ideas of the individual writer. However, in order to maintain coherence in the contributions from various authors, we recommend the following framework:

1 *Title, name of author(s) and affiliation*

2 *Summary*
No more than 100 words

3 *The environment*
Location of the study; the company; the organization
Some key figures about turnover, staff, etc.
Details that might be relevant for placing the study in a proper perspective
How familiar are client and staff with OR support?
How was the contract between client and OR worker established?

4 *The problem*
For which problem did the client call for OR support?
What turned out to be the real problem?
Who was the owner of the problem?
What was the precise assignment?
What did client and consultant expect from the study?
Characteristics: strategic/tactical/operational

5 *The OR approach*
How did the OR consultant handle the problem?
Were any data required? Who collected the data: client or consultant?
What difficulties did the consultant encounter?
How were these difficulties overcome?

Which OR techniques were used to find a solution?
Why was OR used to solve the problem?

6 *Implementation*
How were solutions tested and evaluated?
How were the chosen solutions sold to the organization?
Were any training efforts required?
Did the expected improvements materialize?
Did improvements turn out to be permanent?

7 *Results*
Deliverables for the company/organization
Awareness in the organization
Possible cost savings, once and annual
Possible ideas for improving OR methods

8 *State of affairs*
Conclusions
Recommendations for further action and further implementation
Implementation
Improvements so far
Areas for further investigation, e.g. by academic OR workers

9 *Statement by the company*
Savings and other benefits
Opinion about the OR approach etc.

1.3 The many faces of OR

OR can and should be different things to different people. Any discussion
on a precise definition of OR is bound to be as useful as a debate on the
gender of angels. To some, OR is about developing and applying *tools and
techniques*. To others, it is principally concerned with *problem solving*. Still
others would claim that OR's main preoccupation should be with *structuring
messes*. Who, then, is right? The simple truth is that all of them are!

OR is about developing and applying tools and techniques, about solving
problems and about structuring messes. We should accept this diversity,
cultivate it and be proud of it. Given this diversity, any attempt at structuring
OR applications is going to be incomplete and debatable. Rich OR
applications cannot easily be placed in well-defined pigeon holes. Nevertheless, any book on OR applications requires a structure to guide the reader.
Our attempt to classify OR applications is presented below.

Decision situations can be given attributes such as scope, level, environment, goal, data set. We define them as follows:

SCOPE: Are we dealing with a single component, a connected
                 subsystem or a complex system?

LEVEL: Is the decision situation at the operational level, the
                 tactical level or the strategic level?

ENVIRONMENT: Is it static, evolutionary or truly dynamic?

GOAL: Is the objective well defined and measurable, are
                 there multiple concerns held by different players or
                 are the goals fuzzy, unclear, contradictory and
                 political?

DATA SET: Are we dealing with a small and locally available data
                 set, are data requirements large and to be obtained
                 from different sources, or do we have to deal with
                 huge, noisy and partially incomplete databases?

These attributes can be put in a table that can then be used to classify and
discuss the OR applications in this book (Table 1.2).

We distinguish five parts of OR applications:

1 CLEAN-ROOM OR
2 COMMODITY OR
3 COMPUTER-INTERACTIVE OR
4 FACILITATOR OR
5 CAPSTONE OR

These parts are based on the view of OR as a mature discipline. We will use
this argument to present informally the logic behind our five parts.[2]

Table 1.2 From applying techniques to solving problems to structuring messes (after Ackoff, 1973)

SCOPE	Single component	Connected subsystem	Complex system
LEVEL	Operational decisions	Tactical plans	Strategic scenarios
ENVIRONMENT	Static	Evolutionary	Dynamic
GOAL	Well defined	Multiple concerns	Fuzzy
DATA SET	Small	Large	Huge

When disciplines mature, it is natural that what were once its crown jewels (e.g. linear programming) become accepted and integrated as basic tools in other disciplines (much like calculus, for example). Indeed, linear programming (LP) is quickly becoming a *commodity*, to be activated by pressing a function key in a spreadsheet program. Standard resource allocation problems (maximizing contribution or minimizing cost subject to constraints) are no longer the province of OR specialists. They have become what we call our 'commodity OR' part. Any person with a reasonable education and familiar with spreadsheets can solve a resource allocation problem with LP. See Figure 1.2.

Another natural consequence of a maturing discipline is its increasing tendency towards depth and specialization. It (fortunately!) continuously develops more sophisticated tools and techniques and moves closer to applied mathematics and computer science. This tendency lies at the heart of the so-called 'crisis' debate about the (perceived) gap between theory and practice. There is, of course, a risk that a discipline would become too theoretical and too remote from its practical origins. On the other hand, the trend of 'commoditization' requires a continual development of new tools for the discipline to survive and thrive.

Standard OR tools developed by applied mathematicians and computer scientists working in OR have been successfully embedded in software and hardware systems. There are many examples from combinatorial optimization, such as shortest-path or travelling salesman algorithms in chip design. How long will it be before some of these tools are hardwired in a chip themselves? This is what we call our 'clean-room OR' part where clean room refers not only to a successful application domain but also to the fact that these tools are often developed in a (computer) laboratory with very few (if any) contacts with eventual users. However successful, these applications of OR bear the risk of being 'hidden' to the user. It is indeed difficult for a user to appreciate the power of OR when the user has no idea that sophisticated combinatorial optimization algorithms are a major factor in the efficiency of the user's automated warehouse or flexible machining centre. (See Figure 1.3.)

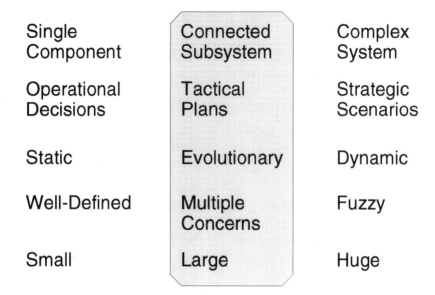

Single Component	Connected Subsystem	Complex System
Operational Decisions	Tactical Plans	Strategic Scenarios
Static	Evolutionary	Dynamic
Well-Defined	Multiple Concerns	Fuzzy
Small	Large	Huge

Figure 1.2   Part 2 Commodity OR.

Pessimists would claim that OR has lost its personality as a result of its own success. It has become either 'commodity OR' in the public domain (LPs in spreadsheets) or 'clean-room OR' (sophisticated technical tools hidden somewhere in a complex hardware or software system). This view is ridiculous. OR is alive and well. We will give three reasons why.

First, clean-room OR is perfectly respectable in its own right. There is potential here as computers become ever more powerful. Moreover, as stated earlier, this work is a permanent source of new and interesting tools, some of which could very well become OR's crown jewels of tomorrow.

Second, the view of commodities is overly simplistic. LP, for example, remains a fairly sophisticated tool (compared with, say, a spreadsheet) requiring an educated user in order to reap fully the potential benefits of its application. Moreover, when non-OR specialists come to trust LP as a standard (commodity) tool, the chances are that they will soon request more sophisticated (non-commodity) OR applications.

Third, OR has evolved in many other directions and all of these are still going strong. There is a whole cluster of applications where OR tools can (for the time being?) at best be a support and a supplement to the knowledge and experience of the user. We call this cluster 'computer-interactive OR', mainly because user-friendly interfaces are an important way of leveraging the performance of the individual decision maker. (We purposely avoid 'decision support systems' since this term means very different things to different people.) The applications in this part are often at an operational, and sometimes at a tactical, level. (See Figure 1.4.) Another cluster of

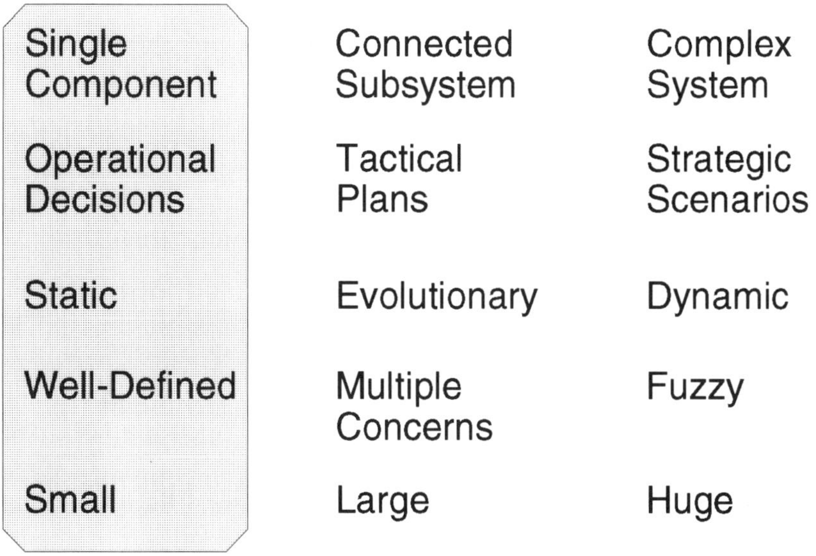

Single Component	Connected Subsystem	Complex System
Operational Decisions	Tactical Plans	Strategic Scenarios
Static	Evolutionary	Dynamic
Well-Defined	Multiple Concerns	Fuzzy
Small	Large	Huge

Figure 1.3 Part 1 Clean-room OR.

applications deals with strategic (sometimes tactical) problems. The support from OR is more on issues of cooperation and integration (cross-functional, multi-plant) and it resides less in the tools or interactive computer systems than in the person of the OR worker and his or her process skills. This is what we term our 'facilitator OR' part (Figure 1.5). Finally, there is a whole

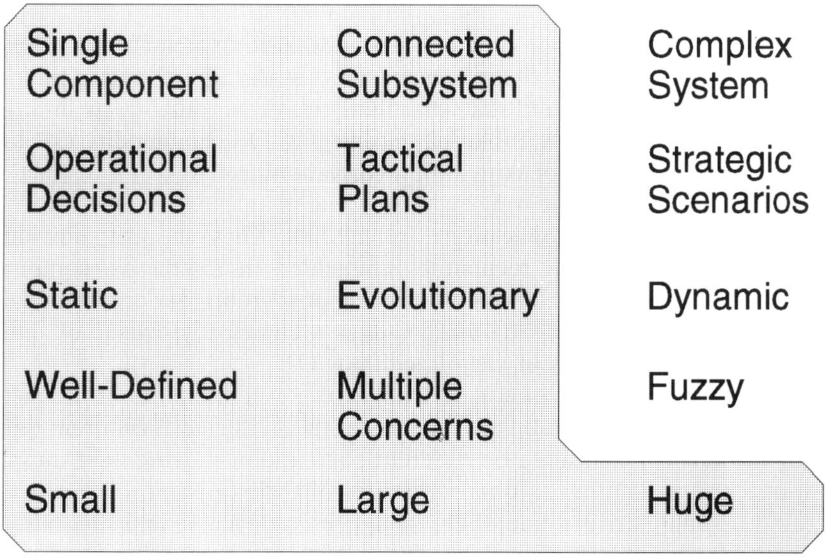

Single Component	Connected Subsystem	Complex System
Operational Decisions	Tactical Plans	Strategic Scenarios
Static	Evolutionary	Dynamic
Well-Defined	Multiple Concerns	Fuzzy
Small	Large	Huge

Figure 1.4 Part 3 Computer-interactive OR.

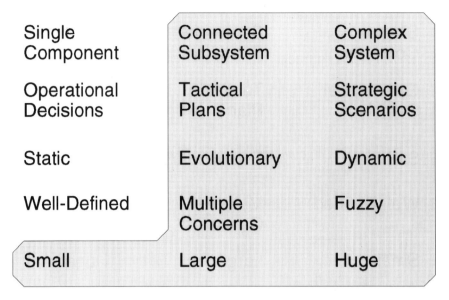

Single Component	Connected Subsystem	Complex System
Operational Decisions	Tactical Plans	Strategic Scenarios
Static	Evolutionary	Dynamic
Well-Defined	Multiple Concerns	Fuzzy
Small	Large	Huge

Figure 1.5   Part 4 Facilitator OR.

set of applications concerned with (scarce) resource allocation on a societal or global level (health management, environmental problems) often dealing with huge, scattered and incomplete databases. This is where OR can support public policy making. Since OR will have to integrate several disciplines, systems and concerns, we have termed this part 'capstone OR'. (Figure 1.6).

There is a certain logic in how one part relates to its neighbours. This is illustrated in Figure 1.7. Looking in more detail at this figure, and at the risk of oversimplifying, one could state that there are three 'pure' parts in terms of the attributes defined in Table 1.2: clean-room OR, commodity OR and capstone OR. Roughly speaking, they correspond to Ackoff's distinction between applying tools, solving problems and managing messes (Ackoff, 1973). However, many applications are hybrids. They support individual decision makers with OR techniques embedded in interactive computer systems (computer-interactive OR) or they support groups of decision makers by acting as facilitators and by using OR as a common language or 'glue' between conflicting points of view (facilitator OR).

Figure 1.7 offers a view of how OR can evolve as a discipline. It has many faces, all of which can be (and are) successful. It also suggests different types of OR workers concentrating on different parts and/or moving their interest from one part to another at different stages of their careers.

Table 1.3 summarizes some of the ideas presented above and suggests yet another perspective. As one moves from Part 1 to Part 5 'process' skills become increasingly important. People issues and a multidisciplinary app-

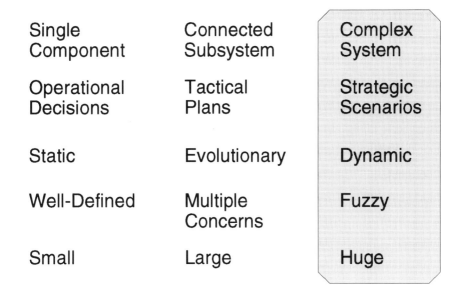

Single Component	Connected Subsystem	Complex System
Operational Decisions	Tactical Plans	Strategic Scenarios
Static	Evolutionary	Dynamic
Well-Defined	Multiple Concerns	Fuzzy
Small	Large	Huge

Figure 1.6   Part 5 Capstone OR.

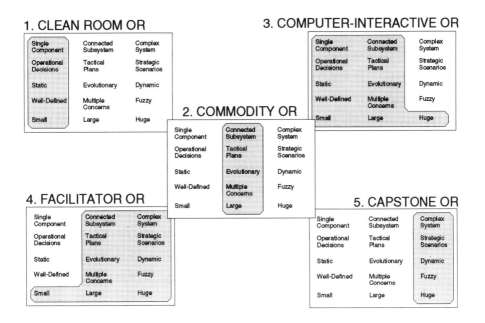

Figure 1.7   Relationship between the five parts.

Table 1.3 OR application parts and worker skill

Part	Process skills	Technical skills	Application type
1. Clean room			hardwired, embedded, hands-off
2. Commodity			semi-automated, impersonal
3. Computer-interactive			person-computer interface
4. Facilitator			group cooperation, business integration
5. Capstone			policy making, societal issues

roach become paramount, a *sine qua non*. Technical expertise, on the other hand, dominates Part 1. However, one should not conclude from this that technical skills are not important to do a good job on the other parts!

1.4 The structure of this book

This book is organized around the five parts: clean-room OR, commodity OR, computer-interactive OR, facilitator OR and capstone OR. However, one should keep in mind that any categorization, although perhaps useful for didactic purposes, has its drawbacks and limitations. It just happens to be the lens through which we look at the complex and rich world of OR applications. You are cordially invited to join us.

We assigned every application in this book to a part. Obviously, some applications exhibit characteristics of several parts (showing the limits of our categorization). In those cases, the application was assigned to the part with which it appeared to have most in common. Many books on applications of OR have a large number of chapters which would fit in our commodity OR part. That is, the applications typically deal with standard resource allocation problems, often using LP as a solution method. A major advantage of our clustering method is that it has allowed us to search for interesting applications in every other part. In so doing, we hope to illustrate better the extreme variety and richness of our discipline.

The application parts are followed by a more reflective part which considers various critical aspects of the discipline of OR. The first chapter in this part traces the origins of the 'crisis debate', i.e. the debate about the (perceived) gap between theory and practice in OR, and conveys some of our personal views on this matter. The second chapter discusses some important issues to be considered when engaging in the process of an OR intervention. This chapter is based on our interviews with OR practitioners and their clients. Together they form a part that we have called 'The process of OR'.

1.5 Parts and chapters

The following list describes the partitioning of this book in application parts and the different applications (chapters) contained in each part. It allows the reader to cherry-pick according to personal preferences.

Part 1   Clean-room OR

Chapter 2   Design of a fast step-and-scan wafer stepper

Summary. This chapter describes the solution of a problem that was passed to the authors by a designer of a wafer stepper, a device used for the photolithographic processing of integrated circuits. In the process of designing such a device, some freedom exists with respect to use of materials, choice of type of motor, and so on. A specific choice implies restrictions on the resulting wafer stepper and its performance, and therefore on the throughput of chips. This is why the authors were asked to develop a tool for determining the maximum throughput of chips, given the mechanical restrictions of the wafer stepper. Their approach consists of two parts. First, they determined the maximum number of chips that can be produced from one silicon disc and, second, they searched for the fastest way in which all chips of such a disc can be processed. Both problems can be solved by combinatorial optimization methods. The resulting algorithms are useful for both the designer as well as the end user of the system.

Chapter 3   Configuration of telephone exchanges

Summary. This chapter describes a project commissioned by AT&T Network System-Netherlands BV and carried out by CQM BV. AT&T-NS produces telephone exchanges. Since each exchange has different traffic characteristics there is no standard and consequently each exchange has to be configured from prefabricated building blocks, the so-called units. These units are placed into switching modules (SMs). In spite of many restrictions which have to be satisfied, there are many ways of doing this. The number of SMs required depends on the placement of units. Hence, an important issue in the exchange design process is the assignment of units to SMs (configuration of the exchange) in such a way that total costs are minimized. In the case described, costs were substantially reduced by using an automated assignment algorithm. This algorithm is embedded in configuration software tools developed by AT&T.

Chapter 4   Stock control of finished goods in a pharmaceutical company

Summary. This chapter addresses the problem of selecting appropriate replenishment rules and ordering parameters for logistics management in the marketing department of a multinational pharmaceutical company. This department has to supply a worldwide net of subsidiaries with finished goods from a centralized production. The policy which is to be implemented should guarantee a high customer service level at low costs of stockholding, production and transhipment. Owing to uncertainties on the demand and replenishment side the problem is tackled by means of stochastic inventory theory. Along with the use of reasonable approximations an easy and near-optimal calculation of control parameters like order-up-to-levels, reorder points and replenishment cycles is possible. Simulation results show that the determined control rules are able to meet prespecified service levels and allow for significant reduction of the relevant costs for most of the analysed products.

Part 2   Commodity OR

Chapter 5   Cutting problems in a production planning environment

Summary. The starting-point of this chapter is the remark that a cutting stock problem can seldom be reduced to a simple standard problem. The examples, given throughout the chapter, should convince the reader that it pays to conduct a careful analysis before implementing a computerized cutting system. The focus is on the analysis and definition of cutting stock problems. A checklist of important issues is provided. The problem definition is taken beyond textbook formulations and hopefully brought closer to what the reader experiences as a real-world problem. General concepts, such as a global cost approach, are introduced and also clarified in a case study. The purpose is not to provide a detailed discussion of mathematical models and solution methods, but to present a general approach to cutting projects. It is stressed that cutting systems remain an intrinsic part of the production planning process and that they should be integrated in the production information system.

Chapter 6   Operational research supports maintenance decision making

Summary. This chapter shows how operational research can support maintenance management. It describes the development and use of two decision support systems which contain maintenance optimization models.

Part 3   Computer-interactive OR

Chapter 7   Shopfloor planning and control for small-batch parts manufacturing

Summary. This chapter describes the development of a planning and control system for discrete parts manufacturing. The main purpose is to illustrate how a series of carefully selected case studies may lead to the development of an OR-based tool, in this case a shopfloor planning and scheduling system. The basic concepts of the system will be illustrated by means of a case study performed at El-O-Matic, a company producing pneumatic and electric actuators for valves. Next, the shopfloor planning and scheduling system is discussed in more detail, with special emphasis on the various resources (machines, cutting tools, pallets, fixtures) that have to be considered simultaneously. Experiments and results with the system at El-O-Matic are briefly reported. The author also indicates how a variety of other cases contributed to the development of the system and concludes with some personal opinions regarding the development of OR-based products.

Chapter 8   Production planning in the fodder industry

Summary. This chapter concerns the development and implementation of a pilot decision support system for a bottleneck in the production facilities of a Dutch fodder-producing company. Every day the planning department of the company is confronted with the problem of determining a time-phased production schedule for multiple products on several, non-identical, parallel production lines, such that the sum of set-up costs and inventory holding costs is minimized. The authors discuss a prototype planning system which has been developed and evaluated in a real-world environment. The test phase showed that the proposed solution techniques are adequate for handling the problem. However, as can be expected during a validation period, they also noticed some shortcomings in the system.

Chapter 9   Operational research methods and their application within the HOT II system of computer-aided planning for public transport

Summary. The use of operational research in public transport goes back to the 1960s. At the time approaches to vehicle and driver scheduling problems were of prime importance. One of the first to use these methods was the Hamburger Hochbahn Aktiengesellschaft, which made its own in-house development of a system (HOT) for computer-aided scheduling. A second-generation system (HOT II) has been successfully implemented by a number of both large and small public transport companies in Europe. The efficiency of the system stems from the OR methods on which it is based, whereby planning procedures are simplified and accelerated, and resultant schedules improved in comparison with manually produced material.

Chapter 10 Planning the size and organization of KLM's aircraft maintenance personnel

Summary. A decision support system (DSS) was developed for the aircraft maintenance department of KLM Royal Dutch Airlines at Schiphol Airport. This department inspects and maintains aircraft during their ground time at the airport. Its main resource is its workforce. Since January 1990 the DSS has supported management in analysing several capacity planning problems related to the size and organization of the workforce. In particular, management uses the DSS to determine the appropriate number of maintenance engineers and their training requirements, and to analyse the efficiency and effectiveness of the maintenance department.

Part 4 Facilitator OR

Chapter 11 Strategic and tactical planning in joint-product industries: a practical application of linear programming

Summary. Joint products are products which cannot be manufactured independently of each other. A typical example is skimmed milk: this cannot be produced without some cream and vice versa. Joint-product industries are those such as the petrochemical, meat, milk and mining industries. Answering strategic and tactical questions with respect to product mix, make or buy policy, commercial approach, production capacities and other resources is especially difficult in these types of industries because of the interdependencies between different products. These interdependencies also make cost calculations difficult. This chapter explains how mathematical programming can be used to provide management with answers to the above-mentioned questions. Practical examples from the milk, meat and mining industries are given to illustrate the approach.

Chapter 12 The use of simulation in the design of a production system for liquid crystal displays

Summary. This chapter discusses the contribution of discrete simulation in the design of a production system for liquid crystal displays. It is one of the few examples in which simulation was used at a very early stage of the design. A project team of a large components manufacturer was responsible for the complete design and installation of the production system. This was a rather large and complex task because both technology and product were relatively new. The simulation project, which was only a very small part of the total project, extended over several years. The author demonstrates that the main contributions of the simulation were on the one hand insight into the logistic design principles of the production system, and on the other hand something best described as 'future reality'. It gave the project team an idea of the final goal of their project by a simple visualization of the working production system.

Chapter 13   Decision support of asset–liability management

Summary. This chapter concerns asset–liability management (ALM) of pension funds. ALM aims to carry out a coherent investment and funding policy, serving the (possibly conflicting) objectives of minimizing the contributions, the fluctuation of the contributions, and the probability of deficits. A primary issue of ALM is the amount of risky securities in the strategic asset mix: if the amount of stocks is increased at the expense of fixed-income securities, a higher expected return will be obtained, which can be used to reduce the contributions, at the expense of a higher volatility of the returns. This in turn will decrease the stability of the contribution rates and the probability of deficits. A second issue of ALM is the sensitivity to inflation of the pension funds' liabilities, and the selection of investments which hedge this inflation risk. In 1985 a decision support system for ALM was developed; it has been extended and improved ever since. This chapter describes the strategic ALM decision problem, the simulation and optimization models which are implemented in the system, some real-life results, as well as the part to be played by the operational researcher in successfully acquiring and carrying out ALM projects.

Chapter 14   A chain model for potted plants

Summary. This chapter shows how OR models of a supply chain can be used to show that reshuffling activities in a distribution chain can substantially reduce costs. As an example, the authors consider the potted plant distribution chain. They use binary programming to model the logical relations in the supply chain. The scenarios they obtain can sometimes be counterintuitive at first sight and may not be obvious without the help of integrated OR models. As such, this chapter demonstrates the usefulness of modelling supply chain issues.

Part 5   Capstone OR

Chapter 15   Reoperations in patients with possibly defective artificial heart valves: the Björk–Shiley dilemma

Summary. Some types of mechanical heart valves (Björk–Shiley convexo-concave valves) have a risk of mechanical failure. So one might consider prophylactic replacement as a preventive measure to avert the disastrous consequences of these failures. A discrete-time Markov model was used to estimate the effect that prophylactic replacement has on survival of individual patients and on medical costs. Model parameters were derived from a large epidemiological study conducted in The Netherlands. For the various valve types, age thresholds were determined, below which replacement prolongs the (discounted and quality-adjusted) life-expectancy. The cost per life year gained as a function of age was also computed. The results of this analysis are used to identify those patients for whom prophylactic replacement might be beneficial.

Chapter 16   OR and the environment: a fruitful combination

Summary. Operational research is a discipline primarily aimed at the improvement of effectiveness and efficiency of decision processes. Those decision processes can be found everywhere in society: in industry, banking, agriculture, government, politics etc. A characteristic of OR is the frequent use of optimization techniques. Since the beginning of the 1980s, these techniques have become more and more user friendly, wrapped up in so-called decision support systems (DSS). This chapter illustrates the significance of OR for the description and solution of environmental problems by looking at problems on which the media report almost daily. The aim of this chapter is to show that OR can play an important role in the new top-priority issue called 'the environment'.

Part 6   The Process of OR

Chapter 17: Crisis? What crisis? Four decades of debate on operational research

Summary. Over the years, the operational research/management science (OR/MS) literature has shown a growing interest in the history of this field, and also a growing concern about its future, even referring to what is called the 'current crisis in OR/MS'. The apparent drop in attention paid by management journals such as the *Harvard Business Review* to OR/MS can be seen as a symptom of that crisis. A large number of articles have been published arguing what might be the causes and consequences of and remedies for the crisis. From these articles, a number of key issues emerge: (1) tool orientation versus problem orientation; (2) client relations; (3) the interdisciplinary nature of OR/MS; (4) the relevance of OR/MS at a strategic level; and (5) the learning effect of an OR/MS study. This chapter provides a brief overview of the crisis debate, emphasizing these five key issues.

Chapter 18   What the cases don't tell us

Summary. This chapter reports on an investigation among OR consultants and their clients. With preference, consultants were selected from those involved in the case studies presented in the foregoing chapters. What were the experiences of a client in the course of an OR consultancy project? What were the difficult moments and how did the consultant handle them? What circumstances, factors, etc., made the project successful? What can the OR consultant learn from this investigation? How can this knowledge be applied in the relationship with clients? These are the kinds of questions addressed, discussed and answered.

The main text of the book concludes with an epilogue (Chapter 19).

Notes

1. The origin of operational research, to be distinguished from technical research, lies in military organizations before and during World War II: first the Royal Air Force preparing for the 'Battle of Britain' and later on the US Navy fighting German U-boats.
2. Our clustering of the chapters is meant to be thought stimulating as well as practical in organizing the book. It is not to be interpreted as a definitive or limitative classification!

References

ACKOFF, R. L. (1973) Science in the systems age: beyond IE, OR and MS, *Operations Research*, **21**, 661–71.

FORTUIN, L., VAN BEEK, P. and VAN WASSENHOVE, L. N. (1992) Operational research can do more for managers than they think!, *OR Insight*, **5**, 1 (January-March), 3–8.

PART ONE

Clean-room OR

Single Component	Connected Subsystem	Complex System
Operational Decisions	Tactical Plans	Strategic Scenarios
Static	Evolutionary	Dynamic
Well-Defined	Multiple Concerns	Fuzzy
Small	Large	Huge

Fingerprints

- A world of highly skilled functional experts; little cross-functional integration and a small 'people' or 'process' content.
- Typically related to operational decision making in a fairly routine, repetitive setting.
- A relatively stable environment not dependent upon last-minute human interaction because of some dynamic event.
- A well-defined goal, e.g. to develop a highly specialized, state-of-the-art, combinatorial optimization algorithm to solve a very specific problem.
- Usually a fairly small data set, e.g. 100 jobs to be scheduled.

Clean-room OR fosters the development of algorithms and techniques that can also be used in other clusters. It has close links with applied mathematics and computer science. A drawback of clean-room OR is its lack of visibility. Successful applications are often hidden in a much bigger system. For instance, successful chip design may depend on good OR algorithms but very few people outside the field of OR are aware of this.

Three examples of clean-room OR are given in this part. The first deals with chip design, the second with configuration of telephone exchanges, and the third with inventory control. Even though these applications are very successful they could easily go unnoticed. Our discipline needs to find ways to market them successfully.

CHAPTER TWO

Design of a fast step-and-scan wafer stepper

C. M. H. KUIJPERS, C. A. J. HURKENS and
J. B. M. MELISSEN

2.1 Introduction

Photolithography plays an important role in the production of integrated circuits (ICs). With this technique patterns are transferred to a disc of semiconducting material (a *wafer*). The optomechanical system which executes this process is referred to as a *wafer stepper* (IEC, 1984).

We consider a special type of wafer stepper that operates according to the *step-and-scan* protocol (Buckley and Karatzas, 1989). An illumination system produces a slit of light which is projected onto a semi-transparent mask, the *reticle*. The light that filters through exposes a small portion of the photoresist-coated wafer (see Figure 2.1). For a correct reticle pattern transference, the reticle has to be passed under the slit of light while at the same time the wafer is moved in the opposite direction. During this scanning phase, both the reticle and the wafer move at prescribed speeds. After the complete reticle has been scanned, the wafer and the reticle are stepped to the positions required for the next exposure. Repositioning of reticle and wafer is done as quickly as possible. The order in which parts of the wafer are scanned and the scanning directions constitute the so-called *scan strategy*. A scan strategy is considered to be optimal if it maximizes the throughput of chips.

Our research focused on the problem of finding near-optimal scan strategies. The request for this study was made by the Philips Research Mechanics Department of Philips Electronics. This department carries out research concerning new technologies for the production of chips at the request of ASM-Lithography, a company which manufactures wafer steppers. Obviously, for ASM-Lithography it is of crucial importance to provide its customers with high-performance machines.

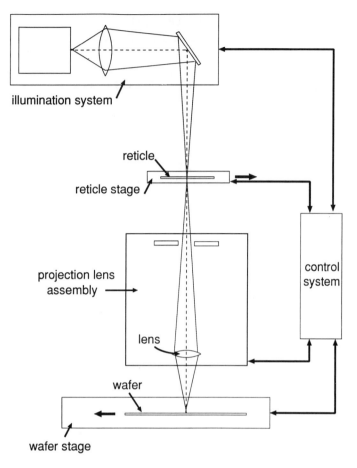

Figure 2.1 Schematic representation of the step-and-scan wafer stepper.

For the designers at the Philips Research Mechanics Department, it was of interest to be able to generate reliable forecasts of the performance of the designed system, given the choice for the parameters ruling the system. Some of these parameters are under the control of the designer, such as size, scale, engine power, properties of optical systems used, and so on. Other parameters may change over time, and are in the hands of the chip manufacturer, such as the size of the wafers and the size of the chips. For the development of a tool that gives such forecasts the Philips Research Mechanics Department sought additional support from the Applied Mathematics Group of Philips Electronics, which in turn contacted the Combinatorial Optimization Group of Eindhoven University of Technology. The result was a project carried out by a student from Eindhoven University of Technology. Good contacts between Philips Electronics and Eindhoven University of Technology have existed for a long time.

In Section 2.2 we split the designers' problem into two combinatorial optimization problems, one of which needs the solution of an optimal control problem. We determine the maximum number of chips that can be placed on a wafer in Section 2.3. In Section 2.4 we formulate the search for optimal movement strategies as a generalized asymmetric travelling salesman problem, which is one of the best known and widely studied combinatorial problems in OR. An algorithm for computing the minimum time between subsequent scans is described in Section 2.5. Each of these sections contains the formulation of the problem, the OR approach for tackling it, and resulting algorithms. A detailed description of our study can be found in Kuijpers (1993) and Kuijpers *et al.* (in press). In Section 2.6 we describe some examples of near-optimal scan strategies. We present our conclusions in Section 2.7, and discuss how the results of our study can be used by the designer, the manufacturer and the user of the wafer stepper.

2.2 Identifying the problems

In designing a scan strategy that maximizes the throughput, one faces two coupled problems. First, the chips have to be placed on the wafer, and second, an optimal scan strategy for this particular arrangement of chips must be found. In theory these two problems have to be considered simultaneously, because it is possible that an arrangement with fewer chips admits a faster scan strategy, which in the end maximizes the total throughput. In practice, however, a large part of the processing time of a wafer is taken up by the loading and the alignment of the wafer. A maximum throughput can therefore only be achieved when the number of chips on the wafer is maximized. Consequently, we can treat the two problems separately. The approach that we take for maximizing the number of chips on a wafer results in an algorithm that is guaranteed to yield an optimal solution. It replaces the heuristic that is currently used.

A configuration of non-overlapping rectangles inside the disc is called a *packing*. Once a particular packing P has been selected, one has to establish the order in which the chips are scanned by the device. Note that the system is able to scan the chips from left to right as well as from right to left. Therefore we also have to decide, for each chip, in which direction it is to be scanned. Each wafer contains two *alignment marks*, which are used for calibration and the exact positioning of the wafer. These marks are etched on the wafer and make the chips in which they are etched unusable. The scanning process must start from one of the alignment marks. A scan strategy is completely determined by the alignment mark from which the scanning process starts, the order in which the chips are scanned and the directions in which they are scanned.

The time required for carrying out the scanning of a given packing P consists of two components. Exposing the wafer to the light is performed on the fly. That is, the wafer moves from left to right, or vice versa, with a constant prescribed speed, and only that part of the chip that is right underneath the lens is exposed. Hence, some time is needed for the actual scanning of the chips. This time does not depend on the scan strategy, since all chips are scanned exactly once, and at a prescribed, constant velocity. Furthermore, some time is taken up by the movements of the wafer and the reticle between the scans of the individual chips. Both the wafer and the reticle need to be at the right positions and have the correct velocities at the start of a scan. The latter time strongly depends on the scan strategy.

Between two subsequent scans there are three independent movements to consider: the movements of the wafer in the x and y directions, respectively, and the movements of the reticle in the x direction. The system can perform these three movements independently and simultaneously. The minimum time needed for these three movements is to be calculated or measured. With this information available, the OR practitioner can recognize an instance of the asymmetric travelling salesman problem. In the classical setting, the travelling salesman problem is the problem of travelling to each of a set of cities exactly once and returning to the city of initial departure, in such a way that the total length of this round trip is minimal. The problem at hand is slightly different, as it is not necessary to make a closed tour. Moreover, the distance is not measured in 'length' but in 'time': starting at some point, 'travel' along all the chips in the minimum total time. Note that, actually, the reticle and the wafer with the chips are moving while the system is stationary. We will see that the times for 'travelling' from the right end point of chip i to the right end point of chip j and the time for the opposite movement are not necessarily the same. Hence, we have to deal with an *asymmetric* travelling salesman problem, with side constraints.

In an early stage of the study, the problems above had already been recognized as such and the idea was to treat them by heuristics. One of the questions for us was whether this heuristic approach would yield answers that were close to optimal. In order to assess the quality of the heuristics one would need to know the true optimal solutions. The heuristics would be used to tackle some exemplary data, in order to figure out the right setting of several parameters governing the mechanics of the system. It turned out that setting up the time–distance table for the travelling salesman problem was much more complicated than expected and that these times depend strongly on the parameters mentioned before. We first needed to solve a rather complex mechanical problem, in order to set up the proper time–distance table.

After identifying the essential problems, we defined the following list of objectives.

- Develop an algorithm for generating all maximum packings.
- Give an algorithm for computing the time for the movement of the wafer stepper between consecutive scans of chips. With this algorithm, construct

a table containing for each pair of scans the minimum time in between.

- Design an effective heuristic for the specific travelling salesman problem and evaluate the feasibility of a combinatorial optimization approach.

- Devise a tool for parameter study by bringing the results together in a decision support system.

Note that the algorithms for finding time-efficient scan strategies are of interest to end-users of the system as well. ASM-Lithography can supply the package as support for its customers. Hence, the results of our work carry tactical as well as operational characteristics.

2.3 Packing chips on a wafer

In this section we consider the problem of finding a maximum packing. Technical requirements restrict us to packings in which the chips are arranged in a rectangular lattice. The corresponding mathematical problem is described as follows. Let W be a truncated disc with radius r in the Euclidean 2-space. A *lattice packing* in W is a collection of rectangles in W each of size Δx by Δy and fitting in a grid. See Figure 2.2 for an example. The maximum number of rectangles of such a packing of W is denoted by $N(W, \Delta x, \Delta y)$.

No simple formulae for $N(W, \Delta x, \Delta y)$ exist, not even for the case when W is a perfect disc. Approximations have been derived (Ferris-Prabhu, 1989). The current approach was based on a heuristic that chooses, within a rectangle of size Δx and Δy, the position of the centre of a circle of radius r, and counts the number of rectangles completely inside the disc. The choices for the centre coordinates were limited to $((k/N)\Delta x, (l/N)\Delta y)$, for $k = 1, \ldots,$ N, $l = 1, \ldots, N$. This procedure is not guaranteed to yield an optimal solution. Actually, this subproblem can be solved to optimality and we devised an algorithm that is guaranteed to generate a maximal lattice packing. The running time is a polynomial of the number of rectangles in the optimal solution. The algorithm is based on the following observation.

Suppose that P is a maximum packing of W. If no lattice point of P touches the border of W, we can move P in any chosen direction until it does. If P touches the border of W in just one point, we can move this point along the border of W until P touches this border in at least two points. Consequently, W has at least one maximum packing that touches the border at least twice.

To find a maximum lattice packing of W we need only consider packings that have at least two points in common with the boundary. The idea behind the algorithm is to generate all these packings. We briefly describe how this is done.

For ease of exposition, let us suppose first that W is a perfect disc. Let P be a lattice packing of W that shares at least two points with the

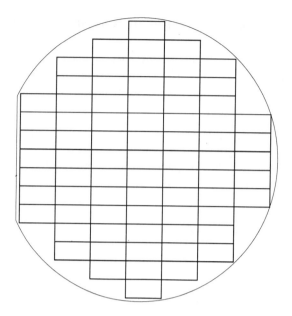

Figure 2.2 A maximum lattice packing on a wafer of radius $r = 100$ mm with rectangles of size 26 mm × 13 mm.

boundary of W, and let the origin O and (p_1, p_2) be these two points. Let M be the centre of W. Since both O and (p_1, p_2) are lattice points, we may write $(p_1, p_2) = (k\Delta x, l\Delta y)$ for some numbers k, l. On the basis of symmetry with respect to the horizontal and vertical axes through M we may assume that k, l are non-negative, and that the coordinates of $M = (m_1, m_2)$ are given by

$$m_1 = \tfrac{1}{2}(k\Delta x - \lambda \cdot l\Delta y) \quad \text{and} \quad m_2 = \tfrac{1}{2}(l\Delta y + \lambda \cdot k\Delta x),$$

with

$$\lambda = \sqrt{\frac{(2r)^2}{(k\Delta x)^2 + (l\Delta y)^2} - 1}.$$

For any choice of k, l, with $(k\Delta x)^2 + (l\Delta y)^2 \leqslant 4r^2$, the centre $M(k, l)$ is then fixed. Let $W(k, l)$ denote the circular disc with centre $M(k, l)$ and radius r, and let $N'(W)$ denote the number of rectangles within W. This number can be easily computed. $N(W, \Delta x, \Delta y)$ is determined by

$$N(W, \Delta x, \Delta y) = \max_{k, l: (k\Delta x)^2 + (l\Delta y)^2 \leqslant 4r^2} N'(W(k, l)).$$

In case W is truncated, the same idea can be used, and only a small modification is needed.

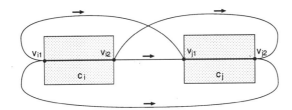

Figure 2.3 Four arcs from chip c_i to chip c_j.

2.4 An asymmetric travelling salesman problem

The problem of determining an optimal scan strategy for a given packing $P = \{c_i | i = 1, 2, \ldots, N\}$ can be converted into a generalized asymmetric travelling salesman problem in the following way.

We represent every chip c_i by two points, v_{i1} and v_{i2}, as shown in Figure 2.3. If v_{i1} is visited before v_{i2}, then chip c_i is scanned in the positive x direction; otherwise it is scanned in the negative x direction. The four arcs in the figure denote the different ways in which chip c_j can be scanned directly after chip c_i. Let v_{01} and v_{02} represent the alignment marks, and v_{N+1} the end point of the scan.

We define a set of *points* $V := \{v_{ij} | i = 0, 1, \ldots, N, j = 1, 2\} \cup \{v_{N+1}\}$, a set of *edges* $E := \{\{v_{i1}, v_{i2}\} | i = 1, 2, \ldots, N\}$, and a set of *arcs* $A := \{(v, w) \mid v, w \in V\}$. The length of an edge is zero; the length of arc (s, t) is defined as the minimum time needed to reposition the system from s to t. The scan problem is now defined as follows: find a path of minimum total length, which starts in v_{01} or in v_{02}, then visits all points of $V \setminus \{v_{01}, v_{02}, v_{N+1}\}$ exactly once, ends in v_{N+1}, and traverses each edge of E. E is referred to as the set of the *obligatory edges*. The problem is actually a mixed rural postman problem (Orloff, 1974; Benavent *et al.*, 1985). We prefer to see it as a travelling salesman problem with the side constraints that some edges have to be on the path. At this point, we assume that the lengths or, better, the 'time distances', are known.

Contrary to the problem described in the previous section we are not able to give an algorithm that solves the stated problem to optimality. Instead, we use a heuristic approach that has proven successful to many combinatorial problems. The general idea behind this is to consider, at any time, some solution to the problem, and try to find a better one by a small modification of the current solution. This principle is known as *local search*. The current solution is locally changed over and over again, until no improvement is likely to be found.

The resulting problem is tackled in the following way. First, a feasible scan strategy (see Figure 2.4) is created. In this figure, the bold line segments represent the obligatory edges. Many methods exist for finding a starting

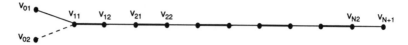

Figure 2.4 Schematic representation of a feasible path.

Random Algorithm	Insertion Algorithm

Random Algorithm

Step 0. Choose at random an alignment mark from which to start.

Step 1. Generate a random permutation π of $\{1, 2, \ldots, N\}$. If $\pi_i = j$, then c_j is the ith chip to be scanned.

Step 2. For every chip, choose a scan direction at random.

(a)

Insertion Algorithm

Step 0. Start with a partial path consisting of one arbitrarily chosen alignment mark v_{0x}, and v_{N+1}. Set $k = 1$.

Step 1. Insert v_{k1} and v_{k2} as cheaply as possible, so that the resulting path represents a feasible scan strategy of the chips $c_1, c_2, \ldots, c_k$.

Step 2. If $k < N$, set $k := k + 1$ and go to step 1; otherwise stop.

(b)

Figure 2.5 Two algorithms for generating a feasible path.

solution (Lawler *et al.*, 1985). In Figure 2.5(a) an algorithm is described for generating a feasible strategy at random. On average, better paths are created with the *insertion* algorithm given in Figure 2.5(b), since it uses information about the cost.

The generated feasible paths are subject to local search algorithms (Lawler *et al.*, 1985; Papadimitriou and Steiglitz, 1982) such as iterative improvement based on 2-exchanges (Croes, 1958) and 3-exchanges (Lin, 1965), and a simulated annealing algorithm (Kirkpatrick *et al.*, 1983; Aarts *et al.*, 1988; Aarts and Korst, 1989) based on 2-exchanges. Here, a 2-exchange refers to the modification of a solution by deleting two arcs from the solution. This splits the path into three parts. It is restored by adding two new arcs and traversing the middle part in the opposite direction (see Figure 2.6). A 3-exchange refers to the obvious analogy with deletion of three arcs. The iterative improvement algorithm only accepts changes that decrease the total path length, whereas the simulated annealing algorithm sometimes also allows changes that increase the total length. The probability of accepting such deteriorations decreases during the execution of the algorithm. The algorithms for improving a scan strategy differ from the standard algorithms in the following. First, the obligatory edges are not eligible for an exchange, as they have to occur in every path. Second, in case the first arc of the feasible

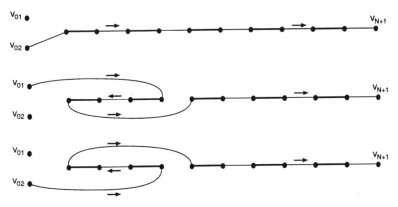

Figure 2.6 Results of 2-exchanges involving the first arc of the path (cf. Figure 2.4).

path is selected for an exchange, it is also possible to change just the starting point of the path.

2.5 Computing the time distances

2.5.1 A one-dimensional mechanical problem

In order to solve the scanning problem we have to be able to calculate the time–distance table. In our case it is not possible to collect the data by measuring the performance of an existing machine or prototype. Remember that our client is the designer who still has to decide upon the several mechanical characteristics of the system. We find an algorithm to compute the time needed between consecutive scans by studying the pure mechanical problem first.

To determine the minimum time required for the movements of the wafer in the x and y directions and for the movement of the reticle, we need to solve three versions of the following one-dimensional problem from mechanics: consider the one-dimensional motion of an object, where $\varphi(t)$ denotes the position of the object at time t. Let $v(t)$ denote the velocity of the object, defined by $v(t) := \dot{\varphi}(t) = d\varphi/dt$. Furthermore, let $a(t) := \dot{v}(t)$ denote the acceleration, and $j(t) := \dot{a}(t)$ the jerk of the object. The mechanical system restricts the ranges of v, a and j by $|v(t)| \leq V$, $|a(t)| \leq A$ and $|j(t)| \leq J$, for all t, where V, A and J are given parameters. What is the minimum time necessary for the object to cover a distance D if its start velocity v_{begin} and end velocity v_{end} are given and uniform, and if at all times $v(t)$, $a(t)$ and $j(t)$ are between prescribed limits? Here, a velocity $v(t)$ is called *uniform* if $a(t) = 0$.

There are two possible scenarios by which the minimum time is achieved:

either the maximum velocity v_{max} reached during the motion is equal to velocity bound V, or v_{max} is smaller than V. Note that at the moment when v_{max} is reached the velocity must be uniform.

If $v_{max} = V$, then the motion can be described as follows. First the velocity is increased as fast as possible from v_{begin} to V. Then the velocity is kept equal to V as long as possible. Finally, as late as possible the velocity is decreased as fast as possible from V to v_{end}.

In case $v_{max} < V$, the motion proceeds as follows: first the velocity is increased as fast as possible from v_{begin} to v_{max} ($v_{begin} \leq v_{max} < V$), after which the velocity is reduced as quickly as possible from v_{max} to v_{end}.

The above observation is the key to solving the problem. What we need to know is the minimum time it takes to accelerate from a uniform velocity v_1 to a uniform velocity v_2, while keeping $a = \dot{v}$ and $j = \dot{a}$ between prescribed limits. Details can be found in Kuijpers (1993). There it is shown that the minimum time $T(v_1, v_1)$ needed to change a uniform motion with velocity v_1 into a uniform motion with velocity v_2 equals

$$T(v_1, v_2) = \begin{cases} |v_2 - v_1|/A + A/J & \text{if } |v_2 - v_1| > A^2/J, \\ 2\sqrt{|v_2 - v_1|/J} & \text{otherwise.} \end{cases}$$

The distance covered in the meantime is

$$x(v_1, v_2) = \begin{cases} \frac{1}{2}(v_2 + v_1)(|v_2 - v_1|/A + A/J) & \text{if } |v_2 - v_1| > A^2/J, \\ (v_2 + v_1)\sqrt{|v_2 - v_1|/J} & \text{otherwise.} \end{cases}$$

With these formulae we solve the inverse problem: given a start velocity v_{begin}, an end velocity v_{end}, and a distance D to cover, what is the minimum time in which this movement can be performed, and what is the maximum velocity attained during this motion?

For D large enough, the maximum velocity is V, and the minimum time necessary is given by

$$T(v_{begin}, V) + T(V, v_{end}) + \frac{D - x(v_{begin}, V) - x(V, v_{end})}{V}.$$

Otherwise, the maximum velocity v_{max} is found by a bisection algorithm that searches for a value of v_{max} satisfying $x(v_{begin}, v_{max}) + x(v_{max}, v_{end}) = D$. In this case the minimum time is given by $T(v_{begin}, v_{max}) + T(v_{max}, v_{end})$.

These results will enable us to compute the minimum time for each of the three movements in between two consecutive scans, since we know, for each pair of chips, the net distance between the end points. Furthermore the start and end velocities are known since they should be equal to the scan velocity. One would expect that the time necessary between two specific scans would be the maximum of the times needed for each of the three independent movements. That this is not always true can be found in Kuijpers (1993) and in Kuijpers *et al.* (in press).

Figure 2.7   The settling times lead to asymmetric distances.

Note that the symmetry of the mechanics problem would lead to a symmetric time–distance table. However, there is one complication that destroys the apparent symmetry. We address this problem in the next subsection.

2.5.2   Settling times

Let us consider more precisely the movements of the wafer and the reticle before a scan. With the results from the previous section we can determine the minimum time needed for the reticle to move to the right position and proceed with a scan velocity $v_{\mathrm{scan},r}$ again. However, at the time the desired velocity is reached, the system is unstable. Some time is required to assure that disturbances and aberrations in the velocity have faded out. This time is called the *settling time* t_r for the reticle. For the same reason, the wafer needs a settling time t_{wx} for the movement in the x direction and t_{wy} for the y direction. These settling times are considered to be given parameters, independent of the prior motions. During its settling time the reticle covers a distance $s_r = |t_r v_{\mathrm{scan},r}|$. Hence, the uniform velocity $v_{\mathrm{scan},r}$ must already be reached at a distance s_r ahead of the position from which the scan starts. Analogously, the wafer has to reach $v_{\mathrm{scan},wx}$ at a distance $s_x = |t_{wx}v_{\mathrm{scan},wx}|$ in the x direction ahead of the position from which the scan starts.

One implication of taking settling times into account is that the distance covered in the motion from chip c_i to chip c_j ($i \neq j$) with scan directions $D(c_i)$ and $D(c_j)$ is in general different from the distance covered in the motion from chip c_j to chip c_i with reversed scan directions (see Figure 2.7). Therefore, the presence of settling times leads to asymmetric time distances.

2.6   Examples

In this section we consider two characteristic examples. Both concern a maximum packing of a wafer with radius $r = 70\,\mathrm{mm}$ with 36 chips of size $20\,\mathrm{mm} \times 15\,\mathrm{mm}$, as shown in Figure 2.8.

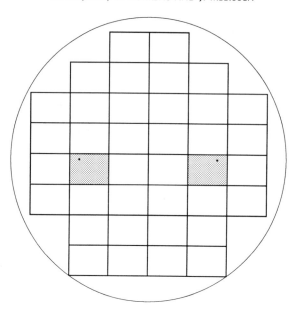

Figure 2.8 An optimal packing. The shaded regions denote chips that are unusable, as they contain the alignment marks.

2.6.1 Example 1

First, we consider the maximum packing shown in Figure 2.8 in combination with the settling times $t_{wx} = t_{wy} = t_r = 0.01$ seconds. From experimentation with different local search strategies we found that the iterative improvement algorithms based on 2- and 3-exchanges outperformed the simulated annealing approach. Improvement algorithms turned out to be essential, because the scans based on random orders were quite bad. Several strategies of length 4.270 360 seconds (scan time not included) were found. One of them is shown in Figure 2.9. Note that the average time in between consecutive scans is approximately 0.12 seconds.

2.6.2 Example 2

We consider the same optimal packing as in Example 1, but now in combination with the following settling times: $t_{wx} = 0.05$, $t_{wy} = 0.01$, $t_r = 0.005$ (seconds). Note that this is a relative extreme case for which the movements in the x direction are relatively time consuming. A zigzag strategy appears to be favourable here (see Figure 2.10).

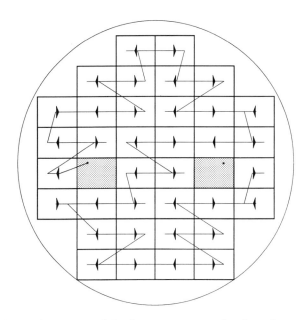

Figure 2.9   Example 1: one of the best scan strategies found.

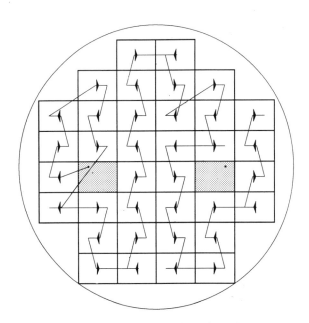

Figure 2.10   Example 2: best scan strategy found, with iterative improvement.

2.7 Conclusions

The problem of determining a fast movement strategy for scanning chips on a wafer has been offered to us in a more or less pure mathematical context. However, the problem was complex in the sense that it was not clear which subproblems were difficult and which ones were easy. Our first goal was to identify the distinct problems and figure out which of these were to be solved to optimality, and which ones we could only approximate.

We decomposed the problem of finding a time-optimal movement strategy for a step-and-scan wafer stepper into three subproblems. The first subproblem was the determination of a maximum packing of a given wafer. This problem has been solved to optimality. The second subproblem was the calculation of the cost (duration) of a given movement strategy. For this, it was necessary to analyse the motions made by the wafer and the reticle between two subsequent scans. We could solve the mechanical problems. Numerical algorithms compute the requested times within arbitrary precision. Finally, the scanning problem was modelled as a travelling salesman problem. We tackled this problem by modifying standard heuristics. Among these heuristics were iterative improvement algorithms based on 2- and 3-exchanges, both combined with multiple starts on random initial paths. Also a simulated annealing algorithm was developed, based on 2 exchanges.

We have conducted a range of experiments not described in this chapter, and from these we could see that the resulting scan strategies indeed depend on various parameters that can be set by the designer of these systems. The scan strategies found with the algorithms described usually contain patterns that are repeated several times. Subsequent chips are always scanned in opposite directions. This can be explained by the amount of time necessary to reposition the reticle in case of two subsequent scans in the same direction. When the settling time of the wafer in the x direction is equal to or larger than the settling time of the wafer in the y direction, we often find that two subsequent scans concern two chips that are situated directly above each other. When the settling time of the wafer in the x direction is considerably smaller than its settling time in the y direction, two subsequent scans often concern chips that are situated next to each other.

The solutions to the various subproblems, and the algorithms for solving them, either to optimality or to an approximate solution, have been delivered to the designer of the wafer stepper. In our research for this project we have made great efforts to find the correct algorithms, and even implement these efficiently. We have delivered building blocks for a proper decision support system, but have not built it ourselves. We left these products in the hands of the designer, who can of course make additional use of our results, by embedding them in a decision support system for the clients of ASM-Lithography. The time frame of our project did not allow for the development of a full-blown decision support system for the designer.

However, our experiments with the available data gave enough insight for a qualitative assessment of the influence of the design on the throughput of the system.

The results are summarized as follows:

1   Given the wafer and chip sizes, both the designer and the end-user of the wafer stepper are able to

■   generate (all) optimal packings.

2   Given a packing and the mechanical restrictions of the wafer stepper, the designer and the end user can

■   generate a feasible scan strategy;

■   evaluate a proposed scan strategy;

■   search for a near-optimal scan strategy;

■   try to improve a proposed scan strategy;

■   study how the throughput and the features of near-optimal scan strategies change when the mechanical characteristics of the wafer stepper are changed – with this, one can evaluate whether it is useful to change the characteristics of the scanner;

■   study how the packing and scan strategy have to be changed in order to obtain maximum throughput when the wafer or chip sizes change.

The setting of parameters clearly influences the lengths of the shortest scan strategies, but the difference between approximately optimal tours is very small. This is the case both relative to the time it takes to handle a wafer, overhead included, as well as compared with the difference between an arbitrary scan strategy and one that is locally optimal, i.e. one subject to our improvement algorithms. Therefore, we conclude that for the design of the system the choice of the parameters ruling the system has a marginal influence on the average performance of the system, expect for rather extreme settings. On the other hand it is clear that, given the set of parameters, it is profitable to use the results to find time-efficient scan strategies.

References

AARTS, E. H. L. and KORST, J. H. M. (1989) *Simulated annealing and Boltzmann machines: A stochastic approach to combinatorial optimization and neural computing*, Chichester: John Wiley, pp. 3–93.

AARTS, E. H. L., KORST, J. H. M. and VAN LAARHOVEN, P. J. M. (1988) A quantitative analysis of the simulated annealing algorithm: A case study for the traveling salesman problem, *Journal of Statistical Physics*, **50**, 187–206.

BENAVENT, E., CAMPOS, V., CORBERÁN, A. and MOTA, E. (1985) Análisis de heurísticos para el Problema del Cartero Rural, *Trabajos de Estadística y de Investigación Operativa*, **36**, 27–38 (in Spanish).

BUCKLEY, J. D. and KARATZAS, C. (1989) Step and scan: A systems overview of a new lithography tool, *SPIE, Optical/Laser Microlithography II*, **1088**, 424–33.

BUSEMANN, H. (1958) *Convex surfaces*, New York: Interscience.

CROES, A. (1958) A method for solving traveling-salesman problems, *Operations Research*, **5**, 791–812.

FERRIS-PRABHU, A. V. (1989) An algebraic expression to count the number of chips on a wafer, *IEEE Circuits and Devices Magazine*, **5**, 37–9.

IEC (1984) *Practical integrated circuit fabrication*, Scottsdale: Integrated Circuit Engineering Corporation.

KIRKPATRICK, S., GELATT, C. D. and VECCHI, M. P. (1983) Optimization by simulated annealing, *Science*, **220**, 671–80.

KUIJPERS, C. M. H. (1993) Determination of a time-optimal movement strategy for a step-and-scan wafer stepper, Master's Thesis, Department of Mathematics and Computing Science, Eindhoven University of Technology.

KUIJPERS, C. M. H., HURKENS, C. A. J. and MELISSEN, J. B. M. Fast movement strategies for a step-and-scan wafer stepper, *Statistica Neerlandica* (in press).

LAWLER, E. L., LENSTRA, J. K., RINNOOY KAN, A. H. G. and SHMOYS, D. B. (eds) (1985) *The traveling salesman problem: A guided tour of combinatorial optimization*, Chichester: John Wiley, pp. 145–60, 215–22, 177–8.

LIN, S. (1965) Computer solutions of the traveling salesman problem, *Bell System Technical Journal*, **44**, 2245–69.

ORLOFF, C. S. (1974) A fundamental problem in vehicle routing, *Networks*, **4**, 35–64.

PAPADIMITRIOU, C. H. and STEIGLITZ, K. (1982) *Combinatorial optimization: Algorithms and complexity*, Englewood Cliffs, NJ: Prentice Hall, pp. 454–86.

CHAPTER THREE

Configuration of telephone exchanges

D. DEN HERTOG, R. W. J. LUKKASSEN and J. F. BENDERS

3.1 The environment

In the 1980s AT&T Network Systems developed the 5ESS switch® telephone exchange, which is a very advanced and flexible digital switching system. It has a modular, distributed architecture consisting of several modules such as a module for administrative purposes, a module for communicative purposes and modules which take care of the actual telephone communication, the switching modules. This type of telephone exchange has been sold all over the world.

Since the demand characteristics differ from customer to customer and from exchange to exchange (traffic load etc.), there is no standard telephone exchange which is suitable for all situations. Each exchange has to be designed from prefabricated building blocks. Each building block performs a specific task in the exchange. Within AT&T, the Office Engineering Systems Development Group is responsible for this design task.

The design process consists of several steps of which two important ones are:

- Determination of the building blocks that are required in the exchange. This depends on the characteristics of the exchange.
- Assignment of the given building blocks to switching modules (SMs), the basic growth module of a 5ESS switch® telephone exchange. Despite the many restrictions that have to be satisfied, there are many ways to assign the building blocks, each resulting in a different number of SMs and different costs. This step is called the *configuration of exchanges* and is the subject of the project described in this chapter.

Design of exchanges is required not only when new exchanges are built but

39

also during the quotation stage. At this stage it is very important that for each new proposal a good (low-cost) exchange can be designed in a short period of time.

The design process is a complicated and time-consuming task which requires a lot of knowledge and experience. This is especially true for large exchanges, which consist of many building blocks. To facilitate and improve this process, AT&T has built a decision support system. This system supports the engineers in the total design process from customer requirements to telephone exchange specifications. It consists of many programs; each program supports a specific step in the process.

One of the programs in the decision support system is developed for assigning the required building blocks to SMs. Doing this so that the total cost of the exchange is minimal is a hard task, since there are a great many ways of doing it, as we will explain in detail later. Although the manually obtained exchange configurations possess very low costs, AT&T realized that even better configurations from the tremendous scale of possibilities could be found by using mathematical tools – more precisely by operational research (OR) techniques. Using a standard software library routine it was able to find optimal configurations for very small telephone exchanges. However, AT&T realized that for larger exchanges it should be possible to find optimal configurations by developing and using special OR methods. Therefore, CQM was asked by this group to develop an algorithm to support the process of automated assignment of the building blocks to SMs.

In the remainder of this chapter the configuration process is described first in more detail in Section 3.2. Next, in Section 3.3, the OR approach to improve the exchanges will be given. Then in Section 3.4 the implementation is discussed. Finally, in Section 3.5 some conclusions and recommendations are given.

3.2 Description of the configuration process

As explained in Section 3.1, the configuration process is that part of the design process of telephone exchanges that is the subject of this project. For this process an algorithm was required that gives optimal configurations. To obtain an insight into the difficulties of the configuration process, a brief outline of this process and a small example will be given in this section.

The building blocks required in the exchange are the starting point of the configuration process. These building blocks are called *units*. The required number of units and their characteristics are a result of the previous step in the design process and therefore known in the configuration process. These units have to be placed into SMs which build up the exchange. Each unit requires a given number of shelves and a given amount of electronic resources such as:

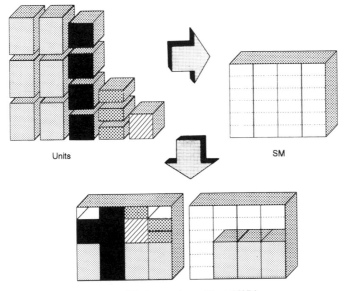

Units   SM

Telephone exchange (several SM's)

Figure 3.1 Assignment of units to SMs.

■ data buses
■ control buses
■ traffic (expressed in time slots)
■ calls per second (processor load).

An SM provides these resources and they are therefore called the *SM resources*. An SM has a defined amount of resources available.

Normally speaking there are several types of units in an exchange. Each unit performs a certain function in the exchange. A set of units that consumes the same amount of resources is called a *unit class*. In practice there may be many types of units per exchange which can lead to up to 60 different unit classes per exchange.

In Figure 3.1 the process of assigning units to SMs is illustrated. In this figure only the SM resource 'physical size' (shelves) is shown. In the upper left of the figure the units to be assigned are given, divided into four unit classes. Each unit requires a certain number of shelves, illustrated by the height of the blocks. In the upper right of the figure an SM is shown. The SM shown has 4×6 shelves available. The configuration task is to put the units into as few SMs as possible such that the available number of shelves (and other resources) per SM is not exceeded. A possible configuration is shown in the lower part of the figure. It is obvious that the units shown fit into two SMs with respect to their physical size. One also has to verify that the other resources consumed by the units in the SM do not exceed the SM capacity.

Apart from the SM resources some units require additional resources which are not provided by the SMs. These resources are provided by special units. If their resources are required by units assigned to the SM, they have to be included into the SM. Such elements are therefore called *forced units*. The resources that are provided by the forced units are called *unit resources*. Just like normal units, forced units require SM resources. We will illustrate this by giving an example:

> Some units need protocol handlers. These protocol handlers are provided by the packet switching unit (PSU): one PSU provides 15 protocol handlers. If a PSU must be placed into the SM, this PSU in turn consumes SM resources (one shelf etc.). If we assign to an SM six units that require three protocol handlers each, two PSUs must be assigned to the SM as well.

The costs of the forced units can be significant.

For an exchange the number of units per unit class that must be in the exchange is given in advance. So, the costs corresponding to these units are fixed, i.e. do not depend on the way they are assigned to the SM. Therefore the costs of the exchange only depend on the required number of SMs and forced units. These numbers depend on the way the units are assigned to the SMs. In general it is possible to save several SMs and forced units by assigning the units wisely. Basically, the aim of the configuration process is to assign the units and the required forced units to SMs in such a way that the costs of the exchange are minimal and for each SM the resources consumed by the units assigned do not exceed the available resources.

Since the costs of SMs and forced units are significant, the quality of the algorithm (costs of the configuration found) is very important. Furthermore, the speed of the algorithm is another important aspect since the algorithm will also be used in the quotation stage (to calculate the costs of the exchange required by the client). A good configuration has to be found in several minutes for average exchange sizes.

In practice it also occurs that an exchange has to be built out of (preconfigured) template SMs, i.e. SMs which are already filled with certain units by the engineers. Then the configuration task is to choose the best mix from these template SMs such that all the units required are in the exchange. The algorithms to be developed should also be able to solve this kind of configuration processes.

This section will conclude by giving an example which will be used in the remainder of this chapter. All figures as well as the SM resources used in this example are not based on real-life telephone exchanges, since these data are confidential.

3.2.1 A small configuration example

In this example we have to assign a number of units belonging to six different classes, respectively 1, 1, 14, 85, 2 and 2 units. Each unit requires some of the SM resources. Each SM provides the following resources: 45 time slots, 32 data buses, 35 control buses, and 24 shelves. These SM resources are listed in the last column of Table 3.1. In the upper left of this table the required amounts of SM resources per unit are given for each unit class. For example, one unit of class 1 (abbreviated to u_1) consumes six time slots, one data bus, two control buses, and one shelf. Moreover, units of class 1 and class 4 require a resource which is provided by forced unit class 1 (abbreviated to fu_1). More specifically:

- One u_1 consumes one of the unit 1 resources; similarly for u_4.
- One fu_1 provides seven unit 1 resources.

Furthermore an fu_1 requires SM resources of two time slots per unit. The cost of one fu_1 is 10 per cent of the cost of a basic SM. In the same way one u_2 and one u_5 require resources that are provided by one fu_2 (three and one respectively). One fu_2 provides six unit two resources. Furthermore one fu_2 requires one shelf and two of unit 1 resource. One fu_2 costs 10 per cent of the basic SM cost.

3.3 The OR approach

In this project the standard OR approach is more or less followed. We will describe the three phases of the standard OR approach (see Figure 3.2) and the way they are implemented in this project. All three phases were carried out in close cooperation with AT&T.

1 Normally an OR consultant starts with a systematic and structured analysis of the subject (Hillier and Lieberman, 1989). The consultant interviews the persons concerned and studies the relevant information to get a clear picture of the process. In Section 3.1 some details of this phase are given.

2 Then a mathematical model is built and solved. In general a model is a representation of a part of the real world. A mathematical model is expressed in the language of mathematics. There may be some gaps in the description (values that are unknown). If the model is formulated so that these values can be calculated then we can solve the model using mathematical techniques. In this project most time was spent in this phase since the techniques that had to be used were not standard OR tools. In Section 3.2 we continue reporting on this phase.

3 Finally the results of the model are translated to the real-life situation and

Table 3.1 An example with six unit classes, two forced unit classes and two unit resource constraints. (Negative numbers indicate that the unit provides the resources, positive numbers that the unit consumes the resources)

Name of unit class		Resource usage of units						Resource usage of forced units		Available per SM
Number of units in the exchange		u_1 1	u_2 1	u_3 14	u_4 85	u_5 2	u_6 2	fu_1	fu_2	
SM resources:	time slots	6	5	4	2	11	0	2	0	45
	data buses	1	2	2	3	2	0	0	0	32
	control buses	2	3	4	1	2	0	0	0	35
	shelves	1	1	1	3	4	4	0	1	24
Unit resources:	unit 1	1	0	0	1	0	0	−7	2	0
	unit 2	0	3	0	0	1	0	0	−6	0

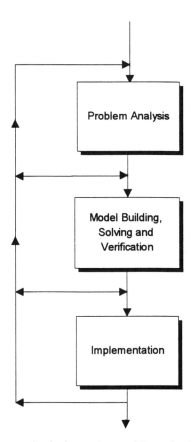

Figure 3.2 The three standard phases in an OR project.

are implemented in cooperation with the client. In Section 3.4 we will discuss this OR consultancy phase in detail for this project.

Of course, there is some feedback between these phases.

3.3.1 Analysis of the configuration process

Before we were able to develop an automated configuration algorithm, a clear picture of the process had to be obtained. In this specific project it was necessary to go rather deeply into the technical details of telephone exchanges. We had to understand the terminology of the engineers to extract the relevant information. The results of this phase are reflected in the previous section. Note that, although that section might seem rather technical, most of the technical details are omitted.

One of the challenges in this phase was to get a clear definition of 'optimal configurations' which was necessary to obtain a good performance criterion for the algorithm. Obviously a first definition would be: exchanges for which the costs are minimal (i.e. there are no other configurations possible that have less costs). However, this kind of 'mathematical problem' is very hard to solve and no algorithms are known that will guarantee to compute the minimal cost configurations within reasonable computation time. So, in general for the larger exchanges it is not possible to check if the solution has minimal costs. Therefore we had to choose a more pragmatic performance criterion. Of course the aim of the algorithm should be to find configurations with minimal costs, but the performance criterion is that the algorithm outperforms manually obtained configurations and that the configurations found could not simply be improved manually. This criterion can be verified and is used to test the quality of the algorithms.

3.3.2 The mathematical model and the solution method

In this section a brief outline of the mathematical model and the solution method is given. Since it is beyond the scope of this book, no detailed description of the algorithm will be given.

In the first phase of the project we obtained a structured view of the configuration process. Using the description in Section 3.2 the derivation of a mathematical model for the configuration process is straightforward. A feasible SM assignment can be described in terms of linear constraints, each SM resource and each unit resource leading to another capacity constraint. So the model can be stated in mathematical terms as:

> A feasible SM assignment must satisfy several (linear) capacity constraints and the objective is to generate a set of feasible SM assignments that contain all units, such that the overall costs are minimal.

The algorithm we developed to find optimal exchange configurations is called the generation–selection algorithm. Here we will discuss the ideas behind this algorithm. As indicated by the name of the method it consists of two parts: the generation method and the selection method. Roughly speaking, in the generation part promising candidate SM assignments are generated, and in the selection part a choice of these candidates is made such that the overall costs are minimal and all units are assigned. The generation–selection framework is summarized in Figure 3.3.

Generation

- Suppose we were able to determine all possible feasible SM assignments. The configuration process would then be reduced to selecting a subset of these SMs such that all units are assigned and the total costs of these SMs

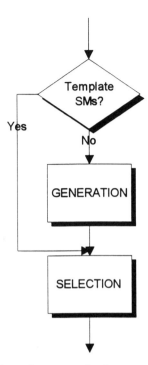

Figure 3.3 Structure of the solution method.

are minimal. However, finding an optimal selection is difficult especially if the number of possible SM assignments is large. Moreover, the number of SM assignments can be so large that it is impossible to generate and store them all. Therefore we want to decrease the number of candidate SM assignments.

- Now the question arises as to which criteria must be used to determine whether a set of generated SMs is sufficient and, if not, which SMs must be added. The so-called generalized linear programming (GLP) approach is used to accomplish this (Gilmore and Gomory, 1961).

- The GLP approach starts with a certain (small) set of SM assignments, and adds other promising SMs. The generation process stops if a set of SMs is found that contains the optimal *non-integer* mix of SMs (e.g. an exchange configuration could be: 1.3 times SM 1, 2.9 times SM 2, etc.). Note that it is not guaranteed that the optimal *integer* mix of SMs is contained in this set. However, experience has shown that only very rarely is the optimal integer mix not in this set. Therefore it is sufficient to start the selection algorithm with this restricted set of SMs instead of all SMs. Depending on the size of the telephone exchange up to 500 candidate SM assignments are produced by the generation algorithm.

Selection

- In the selection algorithm we make a selection from the set of SM assignments produced by the generation algorithm. This problem is known as the multi-dimensional knapsack problem and is carried out using refined branch and bound techniques. The branch and bound technique is a mathematical method to obtain the optimal solution from many possibilities without inspecting all of them explicitly: several subsets of solutions can be discarded at an early stage of the solution process.

In both the generation and the selection algorithm existing OR techniques were used. However, it has to be emphasized that straightforward application of these techniques would not lead to satisfactory results. Therefore the specific structure of the configuration process had to be exploited. In this way both the speed of the algorithm and the quality of the configurations were improved substantially.

Note that if we have to choose from a set of template SMs specified by the engineer, then we can omit the generation algorithm and find configurations directly with the selection algorithm. Hence the generation algorithm can be viewed as a generator of template SMs.

As explained above the candidate SM assignments form an optimal non-integer mix of SMs, and it is not guaranteed that the optimal integer mix of SMs will be contained in this set. Moreover, even if the optimal integer mix is in the set of candidate SMs, to find this optimal integer mix (or a near-optimal one) is often very difficult and the resulting solution cannot always be guaranteed to be the optimal one. However, it is possible to give an upper bound for the percentage of deviation of the configuration found with respect to the optimal configuration. This upper bound is very conservative in practice: normally speaking the obtained configuration is much better than indicated by this upper bound. Although mathematical optimal configurations cannot always be guaranteed, the algorithm provides configurations that cannot easily be improved by hand or simple heuristics.

3.3.3 Example continued: application of the generation–selection algorithm

Let us apply the generation algorithm to the example. This algorithm generates 24 SMs which are listed in Table 3.2. The first number is the SM number, the second the cost of the SM, the following six the numbers of units for the six unit classes, and the last two the numbers of forced units for the two forced unit classes.

The selection method finds the following exchange configuration:

SM 04 (6×); SM 17 (4×); SM 18 (1×); SM 19 (1×).

Table 3.2 SMs generated by the generation–selection algorithm

01	1.2	117202	11	09	1.2	110420	11	17	1.1	003700	10
02	1.2	018002	11	10	1.2	012420	11	18	1.1	002601	10
03	1.1	008301	10	11	1.2	003420	11	19	1.2	110321	11
04	1.2	000800	20	12	1.2	105311	11	20	1.2	011321	11
05	1.2	112122	11	13	1.2	115401	11	21	1.2	010222	11
06	1.2	117202	11	14	1.2	102700	20	22	1.1	101601	10
07	1.2	113212	11	15	1.1	105600	10	23	1.1	100502	10
08	1.2	111221	11	16	1.2	000520	11	24	1.1	001502	10

This configuration needs 12 SMs and 19 forced units, resulting in a cost of 13.9.

3.4 Implementation and results

The generation–selection algorithm has been implemented as a computer program (using the C programming language) and is used within the configuration software tools developed by AT&T.

As input the generation–selection program needs a file with all the required data concerning the resource usage of the SM capacity and the required units. Information about the configuration found is collected in an output file. Note that for the input file it is necessary to understand the mathematical model and to have some experience in translating the telephone exchange data into the model. As a part of the project we also showed the AT&T group how to manage this. In the future AT&T will design a computer program which facilitates the engineers' task of translating the exchange data into the desired (mathematical) format.

To test the generation–selection algorithm a set of test data was provided by AT&T. This was obtained from existing telephone exchanges and therefore realistic tests were possible. The test set consists of exchanges which have from six to 65 different unit classes. In our experience of this limited set of test data we can conclude that good configurations are found by the algorithm. All these configurations satisfy the performance criterion given in Section 3.1 and for some small exchanges it could be proven that the configuration with the lowest costs was found. The configurations are found within 1–10 minutes (on a PC486DX, 33 MHz) for exchanges with up to 40 unit classes. For large exchanges (between 40 and 60 unit classes), good configurations are found within 60 minutes.

For the test set we found an upper bound for the percentage of deviation from optimality of approximately 0–7 per cent. As stated in the previous section this upper bound for the maximum percentage of deviation is very conservative in practice, which means that the performance of the generation–selection algorithm is much better than indicated by the percentages given above. On average, the savings per exchange are two to three SMs for a typical telephone exchange. An SM is very expensive so significant cost reductions can be obtained by using the generation–selection algorithm.

Another important advantage of the developed algorithms, apart from the quality of the obtained configuration, is the speed of obtaining the configuration. Using these algorithms is much faster than calculating a good exchange configuration by inspection based upon the experience and intuition of the engineers. This aspect is of course crucial in the quotation stage, when the costs of the exchange required by the candidate client have to be calculated in a short time, and several alternatives have to be considered.

3.5 Conclusions and recommendations

3.5.1 Conclusions

In the previous sections we have shown how the technical process of configuring telephone exchanges can be solved by using OR methods. We described the three phases in a standard OR project for this specific project. First, we collected the necessary information by carrying out several interviews with the engineers involved. On analysing this information we obtained a good understanding of the configuration process. Second, we built a mathematical model and solved it using advanced OR techniques. These techniques were implemented in a computer program. Finally, the knowledge of how to use the algorithm and how to translate real configuration data into the desired mathematical format was transferred to the AT&T engineers. Using these techniques AT&T is able to offer its clients a better solution regarding its exchanges, which means a competitive advantage.

The algorithms developed were applied to a set of test data. From the results we conclude that medium-sized exchanges can be solved within a short time. It takes more time to solve the large exchanges. For some of the exchanges the configuration found was guaranteed to be the optimal one. It is possible to give an upper bound for this configuration with respect to the maximum deviation of the optimal value. The upper bound is a conservative one.

3.5.2 Conditions for a successful OR application

The configuration process described in this chapter is a typical process which can successfully be solved by OR methods. AT&T satisfied the most important conditions for a successful OR application:

1 *OR awareness*. AT&T realized that OR techniques could be used to optimize their exchange configuration engineering. In many situations, however, one is not aware that OR can offer significant support for complex processes. It is almost trivial, but very important to realize, that OR specialists will not be consulted if organizations are not familiar with the capabilities of OR. Just as 'everybody' knows that one can hire a management consultant to deal with a management problem, in the same way one must know that for certain problems one can consult an OR specialist. In our view it is important for OR specialists to market their capabilities.

2 *OR environment*. OR practitioners have to realize that their solutions have to be embedded within the organizational and/or technical environment of their clients and can only be successful if this environment is ready for the solution. This is a very crucial condition. Concerning the exchange configuration optimization, the AT&T group took care of the organizational and technical embedding of the OR algorithm.

3.5.3 Recommendations for further technical research

Besides the challenge just described, namely making people aware of the capabilities of OR, we highlight the following, related, specific technical areas for further investigation:

- Much research has been done on the so-called knapsack problems (having only one constraint), but only very little on the multi-dimensional variant of this problem, which has many practical applications.
- The generation–selection idea is not new, but we think that this structure (first calculating promising possibilities, then choosing some of them) has been and can be successfully applied to many other problems, e.g. vehicle routing problems or loading problems.

Acknowledgements

We would like to thank the AT&T group for its cooperation in this project and constructive remarks on earlier versions of this chapter. Furthermore we would like to express our gratitude to our colleague, H. A. Fleuren, for his participation in writing this chapter.

References

GILMORE, P. C. and GOMORY, R. E. (1961) A linear programming approach to the cutting stock problem, *Operations Research*, **9**, 849–59.

HILLIER, F. S. and LIEBERMAN, G. J. (1989) *Introduction to Operations Research*, 4th Edn, Oakland, CA: Holden-Day.

CHAPTER FOUR

Stock control of finished goods in a pharmaceutical company

K. INDERFURTH and D. MEIER-BARTHOLD

4.1 Introduction

The main task of the marketing department belonging to a multinational pharmaceutical company is characterized by centrally supplying subsidiaries in more than 100 countries with pharmaceutical products. Replenished with stock from a centralized production this department is responsible for the whole material flow from the final production operations over several intermediate stocking points and transhipments up to final receipt of the goods in the warehouses of the subsidiaries. The marketing department has to manage this replenishment and distribution system connecting sales planning and central production planning.

Owing to the fact that the traditionally employed inventory management system of the company concerned faced considerable problems with respect to amount of stocks, customer service and planning nervousness, the following study was initiated in 1989–90. The main objective of the study was to develop improved inventory control rules providing a prespecified service level while reducing the relevant costs significantly.

In order to understand the specific problems it is necessary to describe the basic situation, which is done in Section 4.2. Thereafter the system is modelled, and detailed requirements for an improved reorder policy are worked out in Section 4.3. Based on these requirements in Section 4.4 two alternative inventory control rules are presented and adjusted to the specific situation. Using simulation the performance of the different control rules is evaluated with respect to two criteria: service levels and relevant costs.

4.2 The planning system under consideration

The traditional planning system, called MDE (Material–Disposition–End-produkte), has been used since 1979. It consists of two consecutive planning processes including (1) sales planning and (2) reorder planning, which are set up step by step every three months. During these runs planning is carried out for nearly 2200–2300 products. This planning procedure is done centrally by the marketing department consulting the subsidiaries and using their data input. The planning horizon is about two years owing to the long production lead times for pharmaceutical products. This technical fact causes a high uncertainty in the input data that can partly be reduced by revising plans in a rolling-horizon planning framework.

4.2.1 Sales planning

Sales planning is a very important part of the overall business planning procedure because it creates the data base for profit planning, sales force targets, and the distribution requirements of the company. Thus, the judgements of the marketing department and the subsidiaries' management supported by statistical forecasting are taken into account. Based on these data the future demand over the whole planning horizon is predicted for each product month by month. These demand forecasts are used as the basis for the whole production and inventory planning system of the company as well as for reorder planning of the marketing department.

4.2.2 Reorder planning

Reorder planning incorporates the centralized stock control for the different subsidiaries in such a manner that on the one hand sufficient service levels at minimum costs are realized and on the other hand convenient production cycles and lot sizes for the final assembly operations (packaging) can be achieved.

The long planning horizon gives rise to several reasons for a revision of reorder plans: (1) changes in sales forecasts, (2) forecasting errors, (3) disturbances on the production side, and (4) transportation delays. Therefore, a flexible planning concept, including successive revisions, called a rolling-horizon framework is used. This concept divides the overall planning horizon into two parts:

1 *Frozen zone.* The frozen zone covers the time consumed by the production planning activities and the physical transportation time. Within this zone a revision of proposed production orders is not allowed because the medium-term activities should not be altered owing to lead

time effects. Therefore, for quick reactions it is necessary to use supplementary short-term activities.

2 *Provisional zone.* The provisional zone of planning periods directly follows the frozen zone. It covers six planning quarters where the provisional medium-term activities are scheduled. Owing to the fact that in each planning cycle input data can be revised each quarter is planned six times.

4.2.2.1 *Medium-term planning*

Based on the sales planning results the medium-term planning of orders for the provisional zone can be performed. These orders are the planned input for central production planning and scheduling operations. In this context the most important quarter is the first one because in the subsequent planning cycle this quarter switches to the frozen zone. Therefore, a further revision due to the medium-term activities cannot be made. Essentially, medium-term planning is based on three strategic control parameters: the minimum inventory level, the order cycle, and the transportation mode.

The minimum inventory level: The main task of the minimum inventory level l_{min} is to protect against the several stochastic influences already described above. In this context the company deserves to realize a very high service level close to 100 per cent. The determination of this strategic control parameter results from a safety cover time (in months) and the average planned demands μ_D per month. In this context the average planned demand is calculated from the average of planned sales over the planning horizon, while the cover time is a prespecified strategic parameter which indicates how many periods of demand have to be covered by the actual stock. In this way the minimum inventory level can be described as follows:

$$\text{minimum inventory level } (l_{min}) = \text{cover time} \times \text{average} \atop \text{planned demand } (\mu_D). \qquad (4.1)$$

The order cycle: Besides the minimum inventory level there is a second strategic control parameter known as the order cycle, which determines the order date and, along with l_{min}, the order quantity. In the company the order cycle is specified by a control parameter called the 'concentration number' K: this control parameter determines the number of aggregated demand periods for a single cycle.

Together with this reorder parameter the medium-term planning is based on a so-called (t,S) replenishment rule with

$\rightarrow$reorder cycle $t =$ 'concentration number' K

$\rightarrow$order up to level $S =$ minimum inventory level (l_{min})
                  $+$ planned demands of K consecutive
                      periods.   (4.2)

The transportation mode: The transportation time in months as defined by the company covers all operations of physical distribution from receipt of finished goods in the central warehouse up to their arrival in the respective warehouses of the subsidiaries. In this context the mode of transport has an important influence on the transportation time. Generally, a distinction has to be made between:

1 Normal mode of transport (e.g. delivery by ship or truck), which is usually employed within medium-term planning.

2 Express mode of transport (e.g. delivery by air), which can be used within short-term planning in case of emergency.

Some problems in medium-term planning: Three important problems in medium-term planning should be mentioned here. They all are caused by the inflexible quarterly procedure of planning even though the planning data are available monthly. Because of the quarterly production planning cycle the reorder cycle K in the company is usually determined by a multiple of 3. If this control parameter differs from that scheme, the internal rule for placing the order date cannot be used. Besides this problem, a convenient production cycle cannot be established because all order due dates are generally fixed to the end of a quarter. Therefore, the necessary production smoothing leads to a premature production and delivery to the central and local warehouses causing an unexpected increase in inventory levels. This means that this so-called 'foredating' of orders by the production department leads to increasing stockholding costs eventually connected by a higher service level. The third problem within medium-term planning is due to quarterly planning on monthly planning data because medium-term planning cannot react immediately to changes in forecasts.

4.2.2.2 Short-term planning

A further problem in medium-term planning arises from forecasting errors within the frozen zone. In this situation the medium-term activities are not able to react in time, which may result in stockouts. To avoid this situation short-term planning combined with short-term activities is needed, which is able to improve customer service provided from the medium-term activities if short-term risks occur. Short-term planning is not treated explicitly in this study but it should be kept in mind that it provides additional protection against uncertainties.

4.3 The traditional reorder policy (model I)

Further considerations will focus on the medium-term reorder policy. This part of the planning process will be modelled and simulated before the requirements for an improved medium-term reorder policy are formulated.

4.3.1 Modelling of the reorder policy

In order to solve a practical inventory control problem a model-based approach appears to be the most systematic way to develop a successful application. However, abstractions from the real-world situation are required to be able to apply models of inventory theory in an appropriate way.

What are the characteristics of the inventory control problem?   Production and distribution of the company's pharmaceutical goods are characterized by multi-level divergent structures. Hence, a multi-stage, multi-product inventory control problem is presented. Owing to organizational reasons the company has separated production planning into single stages. Therefore, the very complex multi-stage, multi-product control problem is reduced to several single-stage ones. Furthermore, the problem is simplified to a set of single-product inventory control problems because the finished goods are specific to individual countries and can be treated separately in the planning system.

How can the uncertainty be modelled appropriately?   Because of the uncertainties of replenishment, transportation and the demand side, which have been described above, a stochastic model is needed. Therefore, considerations are based on theoretical distributions which are fitted to historical data. As a result of statistical tests the normal distribution appears to be appropriate in most cases. Therefore, the normality assumption is used for the calculation of the ordering parameters as well as in the simulation study. The only exception refers to stochastic 'foredating' at the replenishment side which in the simulation study (but not for optimization) is modelled by a triangular distribution which turns out to be very close to the foredating behaviour within a planning quarter.

How can the relevant costs be modelled satisfactorily?   The relevant costs of the medium-term reorder planning consist of reordering and stockholding costs. In the problem under consideration the relevant reordering costs of a product are equal to the set-up costs of a final-stage production run. Stockholding costs are calculated by the value of stock on hand multiplied by the imputed interest rate. For the company it is impossible to determine the relevant shortage costs. This is why the effects of stockouts are taken into consideration by a service level, for which the so-called fill-rate interpretation (β-service level) is used here (see Schneider, 1981; Silver and Peterson, 1985). For medium-term planning transportation costs are not relevant.

Table 4.1   Simulation results for six selected products

	Model I						
	Input data			Calculated data		Output data	
Product	μD	Cover time	K	l_{min}	S	μ_{costs}	μ_β
05	3237	1.5	3	4856	14 567	119.76	99.95
16	148 732	1.0	1	148 732	297 464	438.30	98.21
20	721	2.0	3	1442	3605	1441.18	100.00
29	45	1.2	3	54	189	71.86	96.28
35	7855	1.2	3	9426	32 991	176.71	99.98
36	559	1.0	3	559	2236	147.75	99.83

4.3.2   Performance of the reorder policy

The modelling procedure allows the performance of the medium-term reorder policy to be analysed by a simulation study. The necessary input data for this stochastic simulation include: the estimated means and standard deviations of the uncertain replenishment lead time, transportation time, demand quantities, and the cost of stockholding and reordering. Based on this input and on the company's actual policy parameters, minimum inventory and order up to levels are calculated (see equations (4.1) and (4.2)). The aggregate results of 150 simulation runs with 150 periods per product are shown in Table 4.1 for six selected products out of 35 which have been chosen as representative of different product families.

The main results of the simulation study consist of information on the expected service levels (μ_β) and inventory costs (μ_{costs}). Using these performance measures the replenishment rule and the respective ordering parameters can be assessed. It turns out that the expected β-service level that was not known to the company in advance is usually close to 100 per cent, but for single products significant differences (down to 87 per cent) are observed (see also Figure 4.1).

4.3.3   Requirements for an improved reorder policy

In the practical situation the determination of the ordering parameters is done by the company more or less by using rules of thumb. Thus, it can be expected that a more appropriate determination of the replenishment rule and ordering parameters leads to an improved performance of the system. Therefore, the necessary requirements for an improved medium-term reorder policy can be formulated:

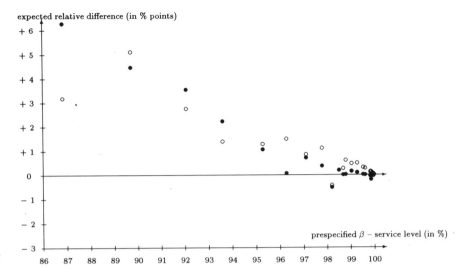

Figure 4.1 Expected relative difference between the prespecified β-service level and the expected β-service level of model II (○) and model III (●).

1   The determination of the replenishment rule and ordering parameters has to be model based allowing applications of OR to optimize the system's behaviour with respect to costs and service levels.

2   The medium-term reorder policy has to take explicitly into consideration uncertainties from the replenishment, transportation and demand side using an optimization procedure.

4.4   Advanced reorder policies

Based on the requirements for an improved reorder policy in this section two alternative models of advanced reorder policies are presented (see Silver and Peterson, 1985):

- *Model II.* A reorder cycle–order-up-to-level replenishment rule with optimized ordering parameters (t^*, S^*).

- *Model III.* A reorder point–order-up-to-level replenishment rule with optimized ordering parameters (s^*, S^*).

4.4.1   Model II: an optimized (t, S) reorder policy

Just as with the traditional reorder policy, model II is based on a reorder cycle–order-up-to-level replenishment rule. The important difference to model I is the fact that the applied ordering parameters (t, S) will be optimized here. Since 'optimization' is used in the sense of cost minimization with respect to the simplified world considered in the underlying inventory model, subsequent simulations have to check how the control rule with model-based parameter determination performs under more realistic conditions.

4.4.1.1 The general optimization problem

Under the following assumptions the optimization of the ordering parameters (t,S) is a well-known problem in the literature: (1) proportional costs of stockholding, (2) fixed set-up costs, (3) fixed replenishment lead times, (4) back order case, (5) only stochastic demand, and (6) stationarity. Under these conditions the stochastic optimization problem can be formulated as follows (see Schneider, 1979):

1 *Objective.* Minimization of the expected inventory costs per period

$$C(t^*, S^*) = \min_{(t,S)} \frac{\sum_{i=1}^{t} L_i(S; h, \lambda) + F}{t} \tag{4.3}$$

where

$$L_i(S; h, \lambda) = h \int_0^S (S - x)\, d\Phi^{i+\lambda}(x).$$

2 *Restriction.* Consideration of a prespecified β-service level

$$\beta = 1 - \frac{\int_S^\infty (x - S)\, \phi^{t+\lambda}(x)\, dx}{t\mu_D} \tag{4.4}$$

with

h unit costs of stockholding per period
F fixed set-up costs
λ lead time
$\Phi^{i+\lambda}$ $(i + \lambda)$-fold convolution of the demand distribution function
$\phi^{t+\lambda}$ $(t + \lambda)$-fold convolution of the demand density function
μ_D expected demand per period.

4.4.1.2 A heuristic for calculating the ordering parameters

This general optimization problem has to be modified for the inventory control problem at hand. This is caused by additional uncertainties on the replenishment and transportation side discussed above. Owing to the complexity of this problem a heuristic procedure is chosen based on the assumption of normality and mutual independence of all stochastic inputs. Hence, it is possible to combine the three distribution functions in an easy way resulting in a normally distributed replenishment lead time demand with parameters $(\mu_{D,\lambda,V}, \sigma_{D,\lambda,V})$ (see Meier-Barthold, 1990).

Now for a prespecified reorder cycle t the order up to level S can be calculated as follows (see Schneider, 1979; Meier-Barthold, 1990):

$$S^* = (t + \mu_\lambda - \mu_V)\mu_D + q_{II}\sigma_{D,\lambda,V}\sqrt{(t + \mu_\lambda - \mu_V)} \tag{4.5}$$

$$\rightarrow l^*_{min} = S^* - (t + \mu_\lambda)\mu_D \tag{4.6}$$

with

μ_λ expected average lead time
μ_V expected average foredating time
q_{II} safety factor.

In this context the safety factor q_{II} is a function of $\beta, \mu_D, \mu_\lambda, \sigma_{D,\lambda,V}$ and t from formulae (4.3) and (4.4). An easy calculation of this function is possible by rational approximation (see Schneider, 1979).

The optimal order up to level has to be calculated by enumeration for each reasonable reorder cycle parameter, which in our case is restricted to a range from 1 to 24 months. Based on the simulation results the performance of the model-optimal (t^*, S^*) policy can be evaluated.

4.4.2 Model III: an optimized (s, S) reorder policy

Another important replenishment rule of stochastic inventory control is the reorder point–order-up-to-level strategy. Contrary to the reorder-cycle-based replenishment rule the time between two reorders is not fixed, thus creating more flexibility in the replenishment system.

4.4.2.1 The general optimization problem

Here, the stochastic optimization problem can be formulated as follows (see Schneider, 1979):

1 *Objective.* Minimization of the expected inventory costs per period

$$C(D^*, s^*) = \min_{(D,s)} \frac{F + L(D + s; h, \lambda) + \int_0^D L(D + s - x; h, \lambda)\,dM(x)}{1 + M(D)} \tag{4.7}$$

where

$$M(x) = \Phi(x) + \int_0^x M(x - r)\,d\Phi(r).$$

2 *Restriction.* Consideration of a prespecified β-service level

$$\beta = 1 - \frac{B(D + s; \lambda) + \int_0^D B(D + s - x; \lambda)\,dM(x)}{[1 + M(D)]\mu_D} \tag{4.8}$$

where

$$B(D + s; \lambda) = \int_{D+s}^{\infty} (x - D - s)\, d\Phi^{1+\lambda}(x)$$

with

D	minimum order quantity ($D = S - s$)
$M(x)$	renewal function
$1 + M(D)$	expected time between two reorders.

4.4.2.2 A heuristic for calculating the ordering parameters

This stochastic inventory control problem is also very complex. Therefore, a simplified heuristic as in model II is needed. For this problem several asymptotic approximations yield an appropriate simplification which allows the ordering parameters (s, S) to be determined step by step (see Schneider, 1978 and 1979).

Determination of the optimal order up to level (S*): Together with asymptotic approximations the minimum order quantity D can generally be determined by a deterministic lot-sizing model. In this situation the economic order quantity is usually used resulting in an optimal order up to level which is calculated as follows:

$$D^* := D_{\mathrm{EOQ}} = \sqrt{\frac{2\mu_D F}{h}}$$
$$\to S^* = s^* + D_{\mathrm{EOQ}}. \tag{4.9}$$

Insufficient approximation is generated by economic order quantities, which are connected with an economic order interval of less than 1.5 periods. In this case the order up to level has to be calculated from model II with the reorder cycle $t = 1$.

Determination of the optimal reorder point (s*) (see Meier-Barthold, 1990):

$$s^* = (\mu_\lambda - \mu_V + 1)\mu_D + q_{\mathrm{III}}\sigma_{D,\lambda,V}\sqrt{\mu_\lambda - \mu_V + 1} \tag{4.10}$$
$$\to l^*_{\min} = s^* - \mu_\lambda \mu_D \tag{4.11}$$

Here the safety factor q_{III} is a function of D, β, μ_D, μ_λ and $\sigma_{D,\lambda,V}$ from formulae (4.7) and (4.8). The calculation of this function can again be supported by a simple rational approximation (see Schneider, 1979).

Table 4.2 Simulation results for six selected products

	Model II					
	Calculated data				Output data	
Product	Cover time	t^*	l^*_{min}	S^*	μ_{costs}	μ_β
05	2.24	9	7235	36 368	79.81	99.99
16	0.64	2	95 718	393 182	337.34	97.77
20	1.57	1	1135	1856	869.25	99.99
29	0.94	13	42	627	33.90	97.78
35	1.91	8	14 981	77 821	137.82	100.00
36	1.36	6	761	4115	131.62	99.95

	Model III					
	Calculated data				Output data	
Product	Cover time	μ_t^*	l^*_{min}	S^*	μ_{costs}	μ_β
05	1.87	9.4	6040	36 406	74.25	99.96
16	0.92	2.4	136 985	501 212	321.45	97.69
20			Results as per model II[a]			
29	0.36	12.8	16	594	31.25	96.35
35	1.76	7.8	13 862	74 780	129.45	99.98
36	1.37	5.8	768	3984	124.43	99.78

[a]See also the explanation in Section 4.4.2.

4.4.3   Performance of model II and model III

4.4.3.1   Some general remarks

In order to determine the optimal reorder cycle t^* from model II each reasonable t parameter has to be considered. Owing to the simulation results it turns out that a global cost minimum exists for an order cycle which is always close to the economic order interval. Therefore, the time-consuming search procedure can be reduced considerably.

As a further important simulation result it should be mentioned that the foredating from the replenishment side has no influence on the optimal reorder cycle t^* and on the optimal expected time between two reorders μ_t^*. Now, applying the alternative models leads to considerable improvements with respect to inventory costs and service levels, which can be seen from the results in Tables 4.1 and 4.2.

4.4.3.2 Comparison with respect to the expected β-service level

In general, the simulation results in Figure 4.1 show that for the policies of model II and model III the prespecified β-service levels, which are chosen according to the simulation results of the actual policy (model I), are achieved and even tend to be exceeded. The reason for this result is the kind of approximations made in the models. An example is the use of a normal distribution as an approximation for the triangular distribution of the time span of foredating when ordering parameters are calculated. This leads to an increasing minimum inventory level caused by the protection against late deliveries. That is why the simulated β-service level increases as well. Furthermore, it can be seen that the expected difference depends on the amount of the prespecified β-service level. In this context model III seems to be the more exact one.

4.4.3.3 Comparison with respect to the expected inventory costs

With respect to cost considerations from our analysis the actual reorder cycle K turned out to be too small in many cases. Figure 4.2 shows the relative cost savings resulting from the application of the advanced planning models. Depending on the difference between the actual reorder cycle K and the optimal one (model II), or the expected time between reorders (model III), the expected cost savings vary by up to 70 per cent. Furthermore, direct comparison of model II and model III indicates average cost savings from 4 to 8 per cent when using model III. Hence, model III, with its more flexible (s, S) control rule, appears to be the better replenishment rule.

Overall, the simulation study shows that for the practical problem under consideration the OR-based determination of the ordering parameters for the originally applied control rule leads to a significant improvement with respect to costs and service, while an additional change of the replenishment rule only results in a slight improvement of the performance.

4.5 Final remarks

The need for this study arose from problems with respect to costs, service, and nervousness caused by the originally applied replenishment rule. As a result of these problems two advanced OR-oriented reorder policies (model II and model III) are presented and evaluated by simulation. Compared with the traditional reorder policy (model I), used by the company, the simulation results show considerable improvements regarding the costs and service levels of the advanced reorder policies.

Based on this simulation study, in 1991 the company's marketing department decided to implement model II for medium-term reorder planning. Since then more and more products have become controlled by this

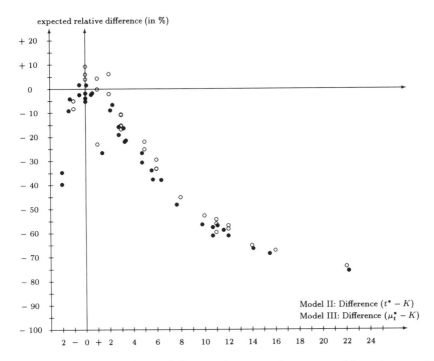

Figure 4.2 Expected relative difference between the expected inventory costs of model I and the expected inventory costs of model II (○) and model III (●).

replenishment rule: nearly 4000 today worldwide. Model II was preferred to model III by the company despite its slightly worse cost performance because it coincided better with the cyclical quarter-based system of production planning. By changing the control rule this system would have to be adjusted as well. Thus, owing to higher-level organizational reasons the company dispensed with the use of a still more cost-efficient control rule. In addition, full optimization of the reorder cycle parameter t in model II has often not been applied because in many cases it conflicted with the quarterly pattern of production set-ups. This means that the demonstrated potential for cost savings in Figure 4.2 could not always be used because of the inflexible planning procedure of the central production system.

Nevertheless, implementation of this advanced medium-term reorder policy combined with a corresponding modified short-term control rule for the frozen zone could improve the traditional inventory management system. Owing to the fact that a prespecified β-service level could be achieved and now observed, safety stocks could be redistributed and reduced. Together with reorder cycle adjustments the evaluated inventory levels could be reduced considerably from DM48 million to DM 27 million (≈ 44 per cent savings) for the German market, for example, while the β-service levels and the nervousness in the planning system considered are equal to the traditional system. Additionally, similar improvements in the logistics performance could be observed for nearly all subsidiaries in more than 100 countries.

References

MEIER-BARTHOLD, D. (1990) Modellgestützte Festlegung von Steuerungsparametern für die Logistik von Fertigprodukten in einem Pharmaunternehmen, University of Bielefeld, Faculty of Economics.

SCHNEIDER, H. (1978) Methods for determining the re-order point of an (s, S) ordering policy when a service level is specified, *Journal of the Operational Research Society*, **29**, 1181–93.

  (1979) *Servicegrade in Lagerhaltungsmodellen*, Berlin: Marchal and Matzenbacher.

  (1981) Effect of service level on order-points or order-levels in inventory models, *International Journal of Production Research*, **19**, 615–31.

SILVER, E. A. and PETERSON, R. (1985) *Decision Systems for Inventory Management and Production Planning*, 2nd Edn, New York: John Wiley.

PART TWO

Commodity OR

Single Component	Connected Subsystem	Complex System
Operational Decisions	Tactical Plans	Strategic Scenarios
Static	Evolutionary	Dynamic
Well-Defined	Multiple Concerns	Fuzzy
Small	Large	Huge

Fingerprints

- Problems are often functional, i.e. the centre of gravity resides in a single department or function in the firm, although inputs are required from other functions and solutions need to be integrated or coordinated with other departments.

- This is the typical textbook type of problem: the tactical medium-term problem of 'optimal allocation of scarce resources', i.e. maximize contribution or minimize cost subject to constraints – the world of LP. A classroom example would be a tactical production planning problem of the multi-product, multi-period, multiple resource type.

- The environment is relatively stable although the data may be constantly updated, e.g. as new orders come in.

- The objective is fairly straightforward although there may be multiple conflicting concerns, e.g. set-up costs versus inventory holding costs, held by different people.

- The data set can be large but is usually manageable, perhaps after some suitable aggregation. There may be tens of thousands of variables and thousands of constraints.

This part shows some of the real traditional power of a model-based OR approach. Large, complex problems can be solved in an environment where human intuition and capability to handle large amounts of data easily fails. Typically, decision makers have a fairly high level of education. Good packaging of the inputs and results is important but so is the quality of the algorithm. Because typical applications have links with several parts of the firm (connected subsystem), the 'process' of OR is important in the design and introduction phases (for acceptance).

The term commodity OR points to the fact that many curricula include some form of introductory OR course, often covering LP. Moreover, LP is now embedded in most spreadsheets and, hence, is readily accessible. However, as the two chapters in this part indicate, this raises the issue of how to make good use of these 'commodity' OR tools! The first chapter deals with an OR classic, i.e. the cutting stock problem, whereas the second discusses the use of OR models in maintenance management. In both chapters the authors argue convincingly that even though the tools of OR may quickly become commodities, their skilful application requires a lot of experience and care. The help of a seasoned OR practitioner is still needed and his or her know-how is by no means a commodity!

CHAPTER FIVE

Cutting problems in a production planning environment

G. SCHEPENS

5.1 In search of the standard cutting stock problem

When managers start their search for a solution to a cutting stock problem, they often look for the standard software package that solves all such problems. Gradually, they discover that most cutting problems have a complex formulation. Even problems related to the same material might be quite different. For instance, software that optimizes the slitting of paper reels will be of no use in a corrugated paperboard plant. Even a new type of slitter might require totally different cutting software.

The successful implementation of a cutting stock problem depends upon management involvement right from the problem definition stage. Management should make a substantial contribution to the effort of getting the problem definition right. For issues related to the choice and implementation of the optimization algorithms, management can rely more fully on outside specialists.

This chapter will therefore focus on the analysis and definition of cutting stock problems. It will take the problem definition beyond textbook formulations and hopefully bring it closer to what the reader experiences as his or her real-world problem. As already stated, cutting problems in this chapter will be different from textbook ones. The underlying assumptions are that:

- the objective should go beyond the usual minimization of waste of raw material;

- the cutting problem cannot be isolated from the production planning system of the plant;

- the problem definition should not ignore the level of technology of the cutting machines (e.g. computer-controlled set-up);

- new business practices, related to just-in-time delivery, result in smaller quantities per order and in less orders per cutting optimization.

5.2   What makes individual cutting problems so different?

The purpose of this section is to draw attention to a multitude of issues that could make a new cutting problem quite different from solved ones. It provides a checklist of those items that should be covered during the analysis:

- the type of material
- the size and shape of the raw material
- the dimensions of the problem
- the shape of the finished product
- the degree of integration within the manufacturing process
- the technology of the cutting machines and the material handling equipment
- the production practices
- the availability of computer control of the cutting equipment
- the crucial impact of commercial requirements.

The basic observation is that a cutting problem can seldom be reduced to a simple 'standard' problem. The examples of this section should prove that it pays to conduct a careful analysis before implementing a computerized cutting system. This approach guarantees that the optimization algorithms will address the right problem. Along the way, management will also discover needs for organizational and technological changes, whose implementation might even exceed the benefits of the computerized cutting system itself.

5.2.1   Type of material

An obvious way of classifying cutting problems is to do it according to the material being cut: glass, paper, corrugated paperboard, solid paperboard, wood, steel, plastic film, foams, textile products, carpets and many more. Even less obvious materials such as diamonds and trees have been tackled by computerized cutting systems.

Materials may have quite specific characteristics:

- Paper can easily be wound up after production, but glass or corrugated paperboard should immediately be cut to size, which requires the integration of the cutting process within the manufacturing process.
- Some materials contain faulty areas, which should be taken into account

during the cutting process. The faults can be of natural origin as in wood, leather or diamond. They can also be due to the production process, e.g. for glass, photographic film or carpets.

- The orientation of the material can be of utmost importance for the strength or appearance of the finished product. Wood cutting should take account of fibre orientation, corrugated paperboard of flute orientation, textile products of weaving orientation.

- The cutting process sometimes relies on the orientation of the material to cleave the product, e.g. diamonds, natural slates.

- Repeated design or colour motifs should not be ignored, e.g. for clothing fabric, carpets or preprint paper.

Although the type of material may look essential, one should not regard it as the single key criterion. There is no single method that solves all the paper cutting problems. On the other hand one will discover astonishing similarities between cutting problems for totally different materials.

5.2.2 Size and shape of the raw material

- The raw material may be available in just one or a variety of sizes. The latter enhances the cutting performance, but will require a higher level of raw-material stock.

- The availability of the raw material in large sizes also improves the cutting performance.

- The shape of the raw material may be important. There is less waste when cutting metal sheets from coils rather than from rectangular sheets of given size.

- Difficult cutting problems may generate large quantities of waste. In some plants, such waste is stored for later use. This 'waste' increases the scope of available raw material sizes, but in practice it might be difficult to keep track of the quantities on hand.

5.2.3 Dimensions of the problem

The number of dimensions relates to the complexity of the problem. Steel rods used in reinforced concrete are undoubtedly three-dimensional objects, but the cutting decisions simply consider one dimension: the rod length. Similarly the cutting of paper reels into smaller reels of the same diameter is regarded as a one-dimensional problem.

The cutting of two-dimensional boards from a larger board with given length and width is a true two-dimensional problem. However, the cutting literature regards some problems as not quite two-dimensional. It classifies

Table 5.1 An example

Dimensions	Examples
One	Rods for reinforced concrete, steel beams, paper reels
Pseudo-two	Sheets cut from reels, e.g. corrugated paperboard
Two	Glass or wood shapes cut from rectangular sheets
Pseudo-three	Extruded beams of foam
Three	Rectangular blocks (marble) Complex three-dimensional shapes (trees, diamonds)

them as pseudo-two-dimensional. An example will be covered in the case study on corrugated paperboard manufacturing. The reason for calling the problem pseudo-two-dimensional is that the base material is a reel for which one dimension can be regarded as infinite (see Table 5.1).

5.2.4 Shape of the finished product

Most computerized cutting problems relate to simple shapes. Most two-dimensional (or pseudo-two-dimensional) problems tackle rectangular shapes. The combination of graphical workstations and powerful interactive software allows irregular shapes to be handled (e.g. in the textile industry or in shipyards).

5.2.5 Degree of integration within the manufacturing process

The key question is whether the manufacturing and cutting process can be uncoupled. In a paper mill this is often the case. The paper leaves the machine on large standard-size reels. The cutting process gets these 'mother reels' from stock, and so operates independently of the paper manufacturing cycle. Of course, management may still want to plan both processes in an integrated way, simply to reduce stocks and throughput time.

Corrugated paperboard or glass cannot be produced on reels; it has to be cut while it leaves the machine. This implies that the cutting problem is embedded in the planning of the manufacturing process. One way to bypass it is to cut standard-size sheets and store them. Customer orders are cut from the stock of rectangular sheets (it is a true two-dimensional problem). Although waste will be higher, the two-stage cutting approach is attractive in the glass industry. It avoids frequent, but short, production runs of each

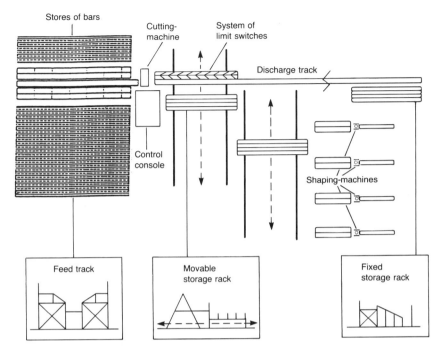

Figure 5.1 Example of an automatic bar cutting and handling system.
Source: Pasche (1991).

raw material. The corrugated paperboard industry cannot afford a large intermediate stock of paperboard, and therefore it limits two-stage cutting to very small orders.

Automated internal transportation systems also affect the cutting process. The automatic feeding of raw material may exclude the use of recycled 'waste' reels, owing to their odd sizes. The automatic transfer and storage of finished products may also introduce extra constraints. For example, modern coil-slitting installations constrain the number of different orders in a set-up, simply because of the limited number of pallet stations at the output end.

An example of an automatic installation, used to cut steel bars, is given in Figure 5.1. The organization and the number of storage bins for finished products affect the amount of feasible cutting patterns.

5.2.6 Technology of the cutting machines and the material handling equipment

The technology of the cutting machine is extremely important. Why should one invest in complex optimization algorithms in order to optimize unadapted and outdated machines? Enhancements to the cutting machines can result in dramatic improvements of the cutting performance. This remark applies particularly to all industrial sectors which face shorter delivery delays and smaller average-order quantities.

Faster and more automated set-ups are a key element in improving cutting performance. Set-up issues are also responsible for much of the complexity of models. Set-up problems occur when loading raw material, when setting the tools for another cutting pattern, or when adjusting the unload side of the machine for other finished products. Some machines are needlessly restricted by a shortage of tools, e.g. a lack of knives. This should be corrected whenever the benefits exceed the extra investment.

Two examples, taken from reel slitting and sheet cutting, will further illustrate the importance of the technology of the cutting machine:

1 Slitters transform 'mother reels' in finished-product reels. The latter are characterized by their width, diameter and core type. Simple slitters wind all final reels on one shaft. Such reels should all be of the same diameter and have similar cores. More sophisticated slitters have two or more shafts, and hence allow mixed patterns with reels of a different diameter or with different cores. Some slitters feature individual winding stations for each reel, in order to guarantee perfect winding tension. The winding of sophisticated plastic film requires such individual stations.

2 Sheet cutting can be accomplished by simple cutting systems, e.g. circular saws that always cut across the sheet. This is called guillotine cutting. Numerically controlled cutting heads allow complex cutting patterns. Figure 5.2 provides an example of different degrees of sophistication in sheet cutting.

5.2.7 Production practices

Each production organization has its own practices. Some are considered essential to avoid confusion or to assure a high standard of quality. Some could be regarded as 'old habits' and should at least be reviewed. If considered relevant, such practices should be part of the computerized cutting system. A few examples follow:

- Two orders are not allowed in the same cutting pattern when their sizes are very close. The fear is that machine operators might get confused when taking the reels from the machine and might ship them to the wrong customer.

- The number of different orders in a cutting pattern is limited in order to keep the paletizing activity simple.

- Some patterns are identified as difficult to produce by the average machine operator, and should therefore be avoided in order to assure high-quality output.

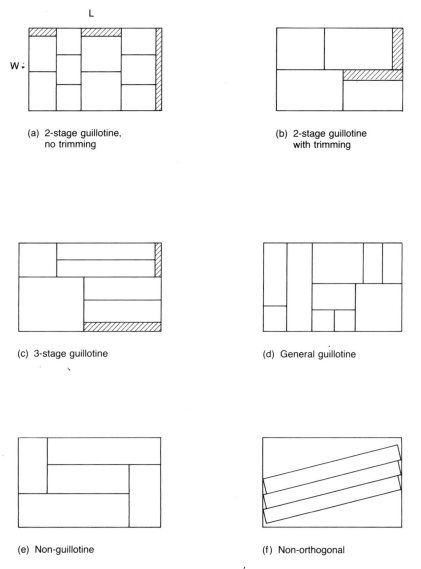

(a) 2-stage guillotine,
    no trimming

(b) 2-stage guillotine
    with trimming

(c) 3-stage guillotine

(d) General guillotine

(e) Non-guillotine

(f) Non-orthogonal

Figure 5.2 Examples of sheet cutting. *Source*: Haessler and Sweeney (1991).

- Patterns containing only one order are kept at the end of the cutting programme. This facilitates quantity corrections in case of rejects.
- When the location of faulty areas is known, narrow finished products (sheets or reels) are positioned in such areas. On average it reduces the amount of rejected output.

5.2.8 Availability of computer control of the cutting equipment

One should also mention the increasing prevalence of computer control systems for cutting machines. They are essential in the race for shorter set-up times. They also make it possible to implement a reliable data interface between the production planning system and the cutting machine. Such interfaces accelerate the downloading of planning decisions and provide accurate feedback on order progress.

5.2.9 The crucial impact of commercial requirements

Commercial business practices really affect the cutting performance. Just-in-time is spreading throughout most of the industrial world. The market has new requirements, as follows:

- smaller order quantities
- shorter lead times
- better delivery performance
- demand for a wider range of raw materials
- production of exact quantities.

Managers are not always aware of the impact of such requirements on cutting performance and some fail to take corrective action, by investing in more flexible and automated cutting equipment.

Smaller-order quantities result in shorter production runs and create set-up conflicts. Shorter lead times and a widening range of raw materials result in fewer orders per cutting optimization, which lowers the average cutting performance. Narrower quantity margins on orders reduce the cutting alternatives. Tricks such as the up- or downgrading of orders to other materials and order-size reductions ('pinching') add alternatives but do not match the total quality image of the supplier.

5.3 Case study: an application taken from the corrugated board industry

5.3.1 The corrugated board industry

The corrugated board industry produces corrugated paperboard, and transforms most of it into packaging products. Corrugated paperboard is made from three, five or seven layers of paper. The even layers (fluting) are corrugated before being glued to the uneven layers (liner).

The paperboard is produced and cut to size on a machine called 'the corrugator' (see Figure 5.3). The yearly production of a corrugator ranges from 40 to 100 million square metres. The average daily production of a plant amounts to 100 orders. A modern corrugated plant will normally have one corrugator and a dozen converting machines. The latter transform the flat boards into boxes, which implies a variety of steps such as printing, die cutting, folding, gluing or stitching. The total investment of a corrugator is approximately BEF400 million (10 million ECU). Corrugated paperboard is a very bulky product. Work in progress in a corrugated-board factory seldom exceeds one day. Both suppliers and customers avoid storing packaging products, which has resulted in smaller order sizes and just-in-time delivery.

5.3.2 Corrugator optimization

The corrugator combines the production of the paperboard (wet end of the corrugator) and its cutting into rectangular sheets (dry end of the corrugator). The cutting optimization on a corrugator must therefore also consider the constraints and costs imposed by the production process of the board. It is the total corrugator that should be optimized. Nowadays, a majority of corrugated board plants use computerized cutting systems, also called deckling systems, to optimize their corrugators.

Let us follow the corrugator downstream, and highlight some of the most important optimization issues:

- The change of paper reels can be quite fast on a modern machine, given the double reel stands and the automatic splicing equipment. The loss of time and paper at each changeover has been greatly reduced, but frequent changes put pressure on the internal transport and mounting of paper reels.

    Some plants operate with several paper widths, and accept that narrow widths induce a loss of machine capacity.

- The full web is first slit into strips by the 'slitter–scorer'. The set-up change of the slitter–scorer is automated on most machines, but remains a cause of paper waste and capacity loss. The two narrow strips at each side of the web are known as the trim. For quality reasons there should always be a minimum trim of a few centimetres, but inefficient cutting patterns may generate high trim wastes.

- A corrugator normally has two (sometimes one or even three) cut-off knives. They chop the strips into sheets. Strips belonging to the same order are guided over the same cut-off knife. The number of knives limits the number of different orders that can be cut in parallel.

- The sheets are stacked and unloaded on automatic conveyors or on pallets. The constraints imposed by this part of the machine are very specific to each site.

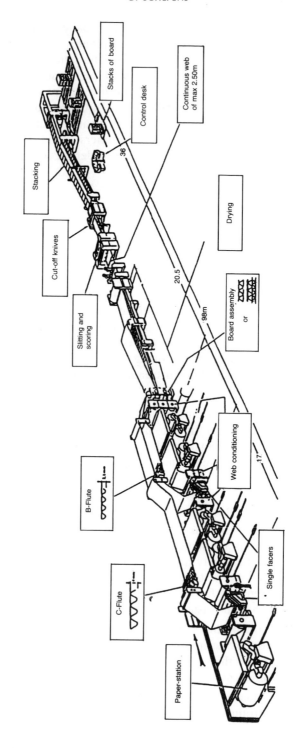

Figure 5.3 Corrugator used to produce and cut corrugated paperboard. *Source:* Adapted and translated from *Emballage Digest*, Ed. Spéciale Cartonneries Associées.

Table 5.2 Board size and ordered quantity

Order	Board width (mm)	Board cut-off length (mm)	Ordered quantity of boards	Allowed surplus quantity
A	1680	2000	625	25
B	460	1000	1180	220
C	360	1500	6333	180

Table 5.3 Required length

Order	Required linear length (m)	Surplus (m)
A	1250	50
B	1180	220
C	9500	270

Trim has long been regarded as the dominant factor commanding attention during the optimization. In this case study, the focus will be first on trim, and then gradually other factors will be introduced.

5.3.3 Numerical data of a sample problem

The sample problem contains three orders. The board sizes and the ordered quantities are given in Table 5.2. The customer requires at least the ordered quantity, but is prepared to accept a limited amount of surplus.

The order quantity can also be expressed in linear metres, as the product of ordered quantity and cut-off length (Table 5.3).

The paper reels are available in four different widths: 2440 (corresponding to the width of the corrugator), 2300, 2200 and 2030 mm.

5.3.4 The concept of cutting pattern

A cutting pattern is a proposal to cut one unit of the raw material. For a one-dimensional problem (e.g. a steel rod), it shows how orders of a given length (order X and Y) could be cut from a base rod (Figure 5.4). The grey area shows the trim.

The corrugated board problem is a pseudo-two-dimensional problem (Figure 5.5). The cutting patterns look precisely as those for a one-dimensional problem. The widths of order X and Y are important to define

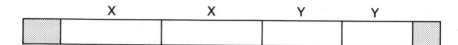

Figure 5.4   One-dimensional pattern.

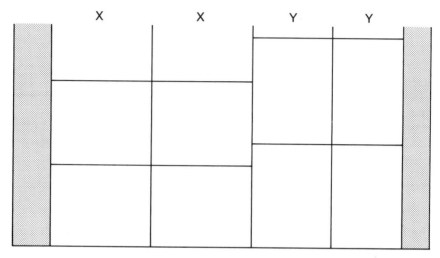

Figure 5.5   Pseudo-two-dimensional pattern.

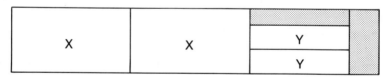

Figure 5.6   Two-dimensional pattern.

the pattern. The cut-off length is not considered here, and the fact that one may lose part of a board at the end of the run is ignored. The cut-off length and the ordered quantity will only be used to compute the production length of the pattern, but that is done during optimization.

A cutting pattern for a true two-dimensional problem, involving order X and Y, can be represented as in Figure 5.6.

5.3.5   Acceptable cutting patterns

A list of acceptable cutting patterns is shown in Table 5.4. This list is not exhaustive. One could imagine producing just order A on a width of 2030 mm at a trim of 350 mm. Here such 'bad' patterns will be ignored, but in practice there are implicit ways to take them into account.

Table 5.4 Acceptable cutting patterns

	Cutting pattern	Effective width (mm)	Width of paper reel (mm)	Trim (mm)	Unused corrugator width (mm)
1	$1 \times 1680 + 1 \times 460$	2140	2200	60	300
2	$1 \times 1680 + 1 \times 360$	2040	2200	160	400
3	$1 \times 1680 + 2 \times 360$	2400	2440	40	40
4	$1 \times\ \ 460 + 4 \times 360$	1900	2030	130	540
5	$1 \times\ \ 460 + 5 \times 360$	2260	2300	40	180
6	$2 \times\ \ 460 + 3 \times 360$	2000	2030	30	440
7	$2 \times\ \ 460 + 4 \times 360$	2360	2440	80	80
8	$3 \times\ \ 460 + 2 \times 360$	2100	2200	100	340
9	$4 \times\ \ 460 + 1 \times 360$	2200	2300	100	240
10	$5 \times\ \ 460\ \ \ \ —$	2300	2440	140	140
11	$6 \times\ \ 360\ \ \ \ —$	2160	2200	40	280

5.3.6 Proposal 1: solution achieving minimal trim

Once the cutting patterns are generated, one can select those that are able to satisfy the demand in an 'optimal way'. In an 'optimal way' depends on the chosen objective. In this first proposal, one requires a solution with minimal side trim.

For each selected cutting pattern, the optimization algorithm will have to specify how much usage will be made from it, i.e. how many linear metres of production (see Table 5.5).

This solution satisfies the demand constraints by producing the required quantity for each order. There is no surplus production. The waste of paper due to trim amounts to 102.6 m^2.

Obviously, all other costs were ignored. No attention was paid to the use of reels whose width is inferior to the corrugator width. The second proposal will consider such evident losses of capacity. All set-up costs related to slitter or reel changes were also ignored, but the third proposal will take set-up changes into account.

5.3.7 Proposal 2: limited cost minimization

The second proposal results from a slightly more sophisticated optimization. The objective function includes:

- the cost due to side trim
- the capacity loss due to partial usage of the corrugator width.

Table 5.5 Solution achieving minimal trim

Width of paper reel (mm)	Cutting pattern	Trim (mm)	Linear length (m)
2440	$1 \times 1680 + 2 \times 360$	40	1250
2200	6×360	40	872
2030	$2 \times 460 + 3 \times 360$	30	590

Table 5.6 Limited cost minimization

Width of paper reel (mm)	Cutting pattern	Trim (mm)	Lost capacity (mm)	Linear length (m)
2440	1×1680 2×360	40	40	1250
2300	1×460 5×360	40	180	1180
2200	6×360	40	280	183

This optimization requires the corrugator speed and some cost data:

Corrugated board: 10 BEF/m^2
Corrugator capacity: 15 000 BEF/h
Corrugator speed: 6000 m/h

The optimal solution is given in Table 5.6. Material wastes, due to side trim, amount to 104.5 m^2, at a total cost of 1045 BEF. The capacity losses on the corrugator amount to 313.6 m^2. If the theoretical capacity is $2.44 \times 6000 = 14\,640$ m^2/h, the cost corresponding to the capacity loss is

$$\frac{313.6}{14\,640} \times 15\,000 = 321 \text{ BEF}$$

On most corrugators, a produced length of 183 m on a given paper width is too short to allow fluent production. A 'minimum run-length constraint' will be added in the next problem formulation.

5.3.8 Proposal 3: extended cost minimization

The third proposal will result from an optimization with an extended objective function. It includes:

- the cost due to side trim

Table 5.7   Extended cost minimization

Width of paper reel (mm)	Cutting pattern	Trim (mm)	Lost capacity (mm)	Linear length (m)
2440	1 × 1680 2 × 360	40	40	1250
2300	1 × 460 5 × 360	40	180	1400

- the capacity loss due to partial use of the corrugator width
- a fixed cost per pattern set-up (due to paper waste and capacity loss)
- a fixed cost per paper reel change (due to paper waste and capacity loss).

Additional constraints are added so that each produced pattern should have a minimal length of 300 m. This optimization requires the corrugator speed and several cost data:

Material cost per m^2 of board: 10 BEF/m^2
Corrugator capacity cost: 15 000 BEF/h
Corrugator speed: 6000 m/h
Set-up cost of a pattern: 350 BEF
Set-up cost at paper reel change: 450 BEF

The optimal solution is given in Table 5.7. The trim increases slightly to 106 m^2, but the use of machine capacity improves and the number of set-ups (both reels and patterns) is reduced by one.

5.3.9   Cost comparison

A comparison of the three proposals is given in Table 5.8. Extra costs in BEF due to the wastes and losses are given in Table 5.9.

Each proposal deserves the label 'optimal', since it is the optimal solution to the corresponding optimization problem. Professionals from the corrugated board industry know that many more model sophistications are required to provide good solutions. But the successive steps of this example illustrate that it pays to go for more elaborate problem formulations.

5.4   Mathematical formulation

As for any other optimization problem, the formulation of a cutting stock problem boils down to the definition of the decision variables, the objective

Table 5.8 Comparison of the three proposals

	Proposal 1	Proposal 2	Proposal 3
Trim (%)	1.68	1.72	1.72
Trim (m^2)	102.6	104.5	106
Capacity loss due to narrow reel width (min)	2.27	1.29	1.24
Number of pattern changes	3	3	2
Number of reel changes	3	3	2

Table 5.9 Extra costs due to waste and losses (in BEF)

	Proposal 1	Proposal 2	Proposal 3
Trim	1026	1045	1060
Capacity loss due to narrow reel width	568	321	310
Pattern changes	1050	1050	700
Reel changes	1350	1350	900
Total	3994	3766	2970

function and the constraints. As stated before, there is no standard formulation valid for all cutting problems. The model in this section corresponds to proposal 3 of the case study, except that set-ups of paper reels are ignored.

Given the following:

I the set of orders
J the set of cutting patterns
L_i the required minimum length of order i
U_i the maximum allowed surplus on order i
M_{ij} the multiplicity of order i in pattern j
A_j the cost due to trim and capacity losses of pattern j (per metre)
P_j the maximum length of pattern j.

5.4.1 Decision variables

- *Which patterns should be used?* These decision variables are represented by the binary variables B_j ($B_j = 1$: pattern j is used; $B_j = 0$: pattern j is not used).

- *How many metres are to be slit according to pattern j?* These decision variables are represented by $X_j \geq 0$. Notice that when pattern j is not used, i.e. B_j equals 0, the run length of the pattern X_j must be zero too. This relation between X_j and B_j will require extra constraints.
- *How many surplus metres are to be produced for order i?* These decision variables are represented by $S_i \geq 0$.

5.4.2 Objective function

The objective is to minimize total costs due to (1) side losses (trim + capacity) and (2) pattern set-up, i.e.

$$\text{minimize} \sum_{j \in J} (A_j X_j + 350 B_j).$$

5.4.3 Constraints

The decision variables must comply with several constraints:

- The produced length equals the sum of the minimum order length and the surplus length:

$$\sum_{j \in J} M_{ij} X_j = L_i + S_i, \quad i \in I.$$

- The surplus length of an order may not exceed the maximum allowed surplus length:

$$S_i \leq U_i, \quad i \in I.$$

- When a pattern is used, it must be run for a minimum length of 300 m:

$$X_j \geq 300 B_j, \quad j \in J.$$

- Whenever a pattern has a positive run length, the corresponding B_j should be one:

$$X_j \leq P_j B_j, \quad j \in J.$$

5.4.4 Extensions

Making an *acceptable* cutting plan requires complying with a range of technical constraints, such as maximum number of cutting knives, and many more. Making a *cost-effective* production plan requires a global minimization of all cost factors, such as various losses of paper, losses of machine capacity, penalties.

5.4.5 Solution algorithms

The models behind Proposals 1 and 2 can be solved with linear programming. All one needs is a generator of cutting patterns and some LP software. The produced lengths of each individual pattern are almost the only decision variables in the problem. The constraints enforce the range of acceptable demand for each order.

The problem behind Proposal 3 requires mixed-integer programming techniques, given that the set-up decisions are modelled by binary variables. In practice, one cannot ignore set-ups, which implies that most real-world problems result in mixed-integer models.

Large sets of orders (e.g. over 30 orders) generate long lists of possible cutting patterns (thousands). So, large-order sets require carefully tuned mixed-integer programming techniques in order to avoid long computation times. (This case study is based on developments done by Beyers Innovative Software. It is a simplified version of a real problem, taken from the corrugated board industry, and solved by OMP. (OMP is a trademark of Beyers Innovative Software NV, Brasschaat, Belgium.))

5.5   A comprehensive approach for cutting stock projects

Section 5.2 focused on the key issues that should be investigated during the analysis of a cutting problem. The case study of Section 5.3 showed that more elaborate problem formulations, matched by appropriate solution methods, increase the quality of the cutting solutions. So, solving cutting problems is clearly more than trim minimization. But the implementation of a cutting stock solver is not a trivial project. This section covers the important steps of a cutting project and also highlights some key issues that should not be overlooked.

5.5.1 Precise definition of the cutting stock problem

At this stage, it should be evident that a good problem definition is the key to success. Section 5.2 was entirely devoted to it. The long list of issues may be used as a checklist during the analysis, although not all of them will be equally relevant for a given project.

5.5.2 The mathematical formulation of the problem

Not all problems will be solved with mathematical programming techniques, but it is always a useful exercise to translate the findings of the analysis into a mathematical model.

Some examples of typical variables are:

- the use made of each cutting pattern;
- set-up decisions related to cutting patterns;
- set-up decisions related to raw material sizes;
- surplus production;
- decisions to upgrade or downgrade an order to a different raw material;
- decisions to produce optional orders (orders with less urgent due date).

Some examples of typical constraints are:

- demand constraints
- technical constraints
  - minimum run length
  - set-up constraints
- availability of raw material (per size).

The objective function should be based on a global cost function, taking into account:

- the various causes of material waste
- the various causes of losses of machine capacity
- some penalties, which act as 'soft constraints'.

5.5.3   The selection of an adequate solution method

Even today, not all problems are optimally solved using mathematical programming techniques. Simple heuristic methods used to be very popular, but they are less justified nowadays owing to the advances made in mathematical programming and the availability of cheap computer power. Continuous linear programming solves a very limited number of real cutting problems. Indeed, set-up decisions introduce binary variables and many problems require integer variables for the use of patterns (e.g. number of reels cut according to a pattern). Therefore, advanced techniques of mathematical programming, blended with artificial intelligence techniques, are an interesting route to follow.

A typical solution strategy is to generate a selected list of cutting patterns first, and then to use mathematical programming techniques to select those that belong in the optimal solution. A method known as 'column generation' creates additional patterns while performing the optimization pivots (mainly applicable with continuous linear programming). The optimization problem may go beyond the selection of cutting patterns; it may also include the optimization of the production sequence. Usually, the cutting optimization and the sequencing of patterns are solved as two separate problems, which holds an evident risk of suboptimization. But it works to embed the sequencing algorithm as a subproblem within the pattern selection optimization.

Problems implying irregular shapes are handled by graphical decision support systems. Some implementations leave most of the decisions to the user, who fits the shapes on the screen based on his or her own insight.

5.5.4 Degrees of freedom

There is no reason to constrain problems more strictly than required. The resulting solution can only be affected adversely. One should therefore preserve the degrees of freedom, whenever this is commercially or technically possible. Some examples of such extra flexibility are:

- tolerance on product sizes
- tolerance on ordered quantity
- tolerance on the use of similar materials
- high flexibility of the cutting equipment
- range of delivery dates, rather than a fixed day
- stocks of popular finished products
- raw material available in several sizes (the price is a higher storage cost).

5.5.5 Integration in the production planning system

Cutting problems normally require that orders, made from the same raw material, be cut together. Large sets of orders give better cutting performance than small ones. So, until recently, it was common practice in most industries to produce orders of a specific raw material not more than once a week. This corresponds to the 'Push' concept, implying high work in process and large finished-goods stocks. Delivery on short notice is made difficult by the fact that each order must join the batch of orders belonging to the same raw material. New concepts such as 'Pull' and 'just-in-time' ask for order launching based on due date. They do not accept that orders are launched in production simply because they are a good match in a cutting problem. But 'order launching based on due date' inevitably results in smaller-order sets, i.e. lower cutting performance. Management can curb the negative effect by, for example, reducing the variety of raw materials and by investing in more flexible cutting machines. Nevertheless the latter implies that management must be aware of the link between shorter lead times and cutting efficiency, and so understands the need to implement a new production approach.

Computerized cutting systems may be used as efficient simulation tools to assess the impact of dramatic reductions of delivery delays and the validity of new production approaches. Planners in charge of cutting machines apply plenty of tricks to lower the waste. Total quality control (TQC) programs have a hard time abolishing such practices as size reduction, overproduction and downgrading of orders to a cheaper raw material. Nowadays, lower cutting efficiencies due to just-in-time make such practices even more popular! Computerized cutting systems are of great help in enforcing TQC requirements related to product specifications, required quantities and due dates.

5.5.6 Integration in the information system

The optimization of the cutting process should not be implemented as a black box. The optimization algorithms should be embedded in a decision support environment. The planners should be allowed to evaluate and fine-tune the proposed solutions.

Computerized cutting systems are no longer standalone applications. They are part of the production planning process and rely heavily on order and production data. The planner should have easy access to complementary unstructured information, e.g. on specific customer requirements. Computerized cutting systems should also have links to the factory floor. Direct downloading of cutting patterns into the control computer of the cutting machine saves time and avoids errors, and on-line feedback on order progress is also very useful.

Cutting performance and compliance with due dates should be monitored. Adequate periodic reports should allow corrective action.

5.5.7 Long-term decisions affect cutting performance

Strategic decisions in the company may have a dramatic impact on cutting performance at the operational level. They set the constraints for the operational problem. Some examples are the following:

- A broadening of the product range (more raw materials) will fragment the orders and result in fewer orders per cutting problem.
- Shorter delivery will have the same effect.
- A lack of flexible, computer-controlled cutting equipment will make the production of small, just-in-time orders both difficult and expensive.

5.6 Expected benefits

Computerized cutting systems should be implemented where the problem is a relevant one. A corrugated paperboard plant that processes BEF1 billion (25 million ECU) worth of paper per year has a relevant problem. When that same plant is pressed to reduce its delivery delays from one week to two days, then the 'relevant problem' is due for an in-depth review.

The traditional benefit of computerized cutting systems is better cutting performance. The aim today is not only lower material waste, but also better usage of machine capacity, less set-ups, reduction of side effects (e.g. order splitting during production). It is not relevant that each individual item is improved; it is the total balance that really matters.

Measurements on cutting performance for recent projects showed a material waste improvement of more than 0.5 per cent. Other operational costs, like capacity costs, normally show more dramatic improvements (e.g. 5 per cent and up), simply because less attention was paid to them before. But the implementation of a computerized cutting system also improves customer service by reducing late deliveries, high production overruns, excessive usage of order upgrades or downgrades. Important benefits also originate from its integration in the information system:

- data are no longer copied manually, which reduces the risk for errors;
- data need to be more accurate in order to make the system work, so people are forced to maintain correct product specifications, accurate raw material tables, up-to-date machine parameters;
- a substantial amount of clerical work can be saved.

The integration of production planning and cutting optimization must result in a higher response to customer demand. Frequent rescheduling is made easier, faster and more reliable.

5.7 Conclusions

For plants with a relevant cutting problem, the question is no longer whether managers should look at computerized cutting optimization, but rather how they should approach it. State-of-the-art cutting systems are part of the production planning system and should be well integrated in the information system of the plant. The impact of changing commercial practices on cutting systems was highlighted. This chapter tried to motivate management to get involved, and it presented hints to define and tackle the real problem.

References

HAESSLER, R. W. and SWEENEY, P. E. (1991) Cutting stock problems and solution procedures, *European Journal of Operational Research*, **54**, 141–50.

PASCHE, C. (1991) Case study: production planning, *European Journal of Operational Research*, **50**, 27–36.

Bibliography

ABEL, D., DYCKHOFF, H., GAL, T. and KRUSE, H.-J. (1983) Klassifizierung realer Verschnittprobleme, *Diskussionsbeitrag*, Nr 66, May.
BEASLEY, J. E. (1985a) Bounds for two-dimensional cutting, *Journal of the Operational Research Society*, **36**, 71–4.
(1985b) An exact two-dimensional non-guillotine cutting tree search procedure, *Operations Research*, **33**, 49–64.
(1985c) Algorithms for unconstrained two-dimensional guillotine cutting, *Journal of the Operational Research Society*, **36**, 297–306.
COFFMAN, E. G. Jr and SHOR, P. W. (1990) Average-case analysis of cutting and packing in two dimensions, *European Journal of Operational Research*, **44**, 134–44.
DAGLI, C. H. (1990) Knowledge-based systems for cutting stock problems, *European Journal of Operational Research*, **44**, 160–6.
DANIELS, J. J. and GHANDFOROUSH, P. (1990) An improved algorithm for the non-guillotine-constrained cutting-stock problem, *Journal of the Operational Research Society*, **41**, 141–9.
DYCKHOFF, H. (1990) A typology of cutting and packing problems, *European Journal of Operational Research*, **44**, 145–59.
FARLEY, A. A. (1988) Mathematical programming models for cutting-stock problems in the clothing industry, *Journal of the Operational Research Society*, **39**, 41–53.
(1990a) The cutting stock problem in the canvas industry, *European Journal of Operational Research*, **44**, 247–55.
(1990b) Selection of stockplate characteristics and cutting style for two dimensional cutting stock situations, *European Journal of Operational Research*, **44**, 239–46.
SOEIRO FERREIRA, J., ANTONIO NEVES, M. and FONSECA E CASTRO, P. (1990) A two-phase roll cutting problem, *European Journal of Operational Research*, **44**, 185–96.
GOULIMIS, C. (1990) Optimal solutions for the cutting stock problem, *European Journal of Operational Research*, **44**, 197–208.
HAESSLER, R. W. (1988) Selection and design of heuristic procedures for solving roll trim problems, *Management Science*, **34**, 1460–71.
OLIVEIRA, J. F. and SOEIRO FERREIRA, J. (1990) An improved version of Wang's algorithm for two-dimensional cutting problems, *European Journal of Operational Research*, **44**, 256–66.
RINNOOY KAN, A. H. G., DE WIT, J. R. and WIJMENGA, R. TH. (1987) Nonorthogonal two-dimensional cutting patterns, *Management Science*, **33**, 670–84.
SCHNEIDER, W. (1988) Trim-loss minimization in a crepe-rubber mill; optimal solution versus heuristic in the 2(3)-dimensional case, *European Journal of Operational Research*, **34**, 273–81.
STADTLER, H. (1990) A one-dimensional cutting stock problem in the aluminium

industry and its solution, *European Journal of Operational Research*, **44**, 209–23.

SWEENEY, P. E. and HAESSLER, R. W. (1990) One-dimensional cutting stock decisions for rolls with multiple quality grades, *European Journal of Operational Research*, **44**, 224–31.

THORSEN, M. N. and VALQUI VIDAL, R. V. (1991) Operational research in the Danish steel industry, *European Journal of Operational Research*, **51**, 301–9.

CHAPTER SIX

Operational research supports maintenance decision making

R. DEKKER and C. VAN RIJN

6.1 The environment: maintenance management

Maintenance can be defined as the combination of all technical and associated administrative actions intended to retain an item or system in, or restore it to, a state in which it can perform its required function (British Standards Institution BS 3811, 1984). Maintenance can pertain to many systems, industrial production systems, civil infrastructures, software and even human beings. Despite some similarities we will not consider the last two in this contribution. Maintenance has long been underrated as a discipline and attracted scientific attention relatively late. Still most mechanical engineering curricula are focused on design of systems rather than on their maintenance. Yet maintenance spending has increased over the years and often constitutes the second-largest cost factor (after energy) in automated industrial plants, like refineries (Solomon, 1989). The maintenance department can be the largest department in staffing! In the civil sector there is also a shift from building new objects to maintaining them. This can be seen not only in the area of roads, but also in buildings, bridges, dams and other infrastructure.

Traditionally, maintenance has been regarded as a cost-making function only, just like the personnel department. A major problem is to determine the annual budget and it is still common practice to use past budgets and correct them for inflation for any changes made. The scientific viewpoint, however, is that maintenance budgets should be determined by the impact of maintenance on system performance and thus on its contribution to company profits. This relation, however, is often difficult to establish, yet essential for the existence of a maintenance department.

Mechanical engineering has significantly contributed to maintenance, both in the area of failure analysis and subsequent system modification to design out maintenance, as well as in the area of measuring devices for predicting the onset of failures (condition-based maintenance). However, it has failed in quantifying and adjusting the contribution of maintenance to company profits. In this chapter we will therefore show how operational research techniques can contribute to maintenance management decision making.

6.2 The problems

To demonstrate the impact of operational research on maintenance we will discuss two industrial cases:

1 *Planning and scheduling of preventive maintenance activities for a major gas turbine.* A gas turbine is a large industrial machine, worth several millions of dollars, which is capable of generating substantial amounts of power. In the case considered one gas turbine served to generate electricity and four others to drive large pumps. They were installed on an offshore oil and production platform which also served as a main pumping station for adjacent platforms. The preventive maintenance consisted of a multitude of activities varying from checks and adjustments of instruments to replacements of individual components like filters. The problem with respect to the execution of these activities was that they required a shutdown of the turbine. Although most platforms have a yearly shutdown for maintenance, this requires that the maintenance interval is at least a year and, moreover, any shortening of the shutdown saves money. During the year there were some occasions on which one of the turbines was shut down for other reasons, like oil-well maintenance. These moments provided opportunities for maintenance, yet they could hardly be predicted in advance and were of limited duration. Hence a prioritization of the work was necessary. At the same time, maintenance management required an assessment of the cost–benefit and optimal interval of all preventive maintenance. To overcome these problems a decision support system (DSS) was developed which will be discussed in this chapter.

   Note that this problem has both operational (which activities to do now) and tactical aspects (what frequency is best in the long run). The problem required support for repeated decision making, i.e. at each opportunity. Three parties were involved: the maintenance planner on board the platform, the onshore supporting staff, and the developers of the DSS. The problem was primarily tackled by the development team, assisted by the supporting staff.

2 *Maintenance of furnace tubes in a refinery.* An industrial furnace is used to heat a stream of fluid before it enters a processing unit. It consists amongst other things of a burner and a number of tubes through which the stream flows. Owing to the high temperature coking will develop in the tubes. As a consequence the process stream is reduced and the

energy efficiency of the furnace is lowered. Moreover, coking will lead to 'hot spots' with subsequently a large probability of tube burst or leakage, requiring an immediate shutdown of the plant. To prevent this the furnace tubes are cleaned regularly, requiring either a shutdown or a reduced throughput of the processing unit it serves. Hence, clean-out costs are quite high. At a certain time the processing unit behind the furnace had to undergo a shutdown and one wondered whether the cleaning of the furnace should be combined with this shutdown. The company in question had a DSS for maintenance optimization in house and the problem was tackled by the engineer responsible for this system. A project team was set up consisting of both maintenance and production personnel to tackle the problem. In this case the problem needed to be solved only once.

6.3 The OR approach

6.3.1 The turbine problem

In the first case we started by structuring the problem, studying the literature and investigating possible solutions. This lasted for about half a year. The problem was quite new, there were no commercial DSSs capable of doing the task (apart from a forerunner developed by former colleagues; see Van Aken et al., 1984), neither was there any literature on the priority setting of maintenance activities. Besides, the original question was to find out how much maintenance is cost effective. Since data acquisition would be a major task and the project team had to come up with results within a year, it was decided to develop some models first, next to build a prototype DSS and then to start with large-scale data collection. The motivation was that we had to show something first before we could involve the supporting staff in large data collection activities.

The DSS developed previously produced a prioritization of maintenance activities based on their contribution to system reliability. Contrary to the original design objectives, this DSS did not function properly in practice. By better structuring of the maintenance activities the associated workloads became considerably easier. Hence more opportunities of limited duration could be exploited. Since correct preventive maintenance always contributes to system reliability the onshore staff had to monitor the process to prevent overmaintenance. In the selection for a better optimization criterion, system availability was rejected. The economics department at that time had difficulty in attributing an economic value to system availability. For the DSS developers an associated problem arose that would introduce dependencies between maintenance activities. Hence it was decided to use total average costs as the criterion, including both direct and indirect maintenance costs (the last due to downtime).

The first problem was that management had little idea of the costs of downtime at unit level (like a gas turbine), since there was some redundancy. Even at system level (like water injection) defining an economic penalty was difficult, because production was deferred rather than lost, because of a complex tax regime and because the variety of actions that could be undertaken upon failure aggravated the problem. Nevertheless, the economics department had developed a reference value for the loss of production of one barrel of oil or cubic metre of gas, to be used in comparing projects. We decided to adopt this value. The next problem was to relate this figure, which was at system level, to that of individual units. To this end a special model was formulated. All systems were modelled as subsystems in a series configuration (implying that failure of one would induce system failure), where each subsystem consists of a k-out-of-n configuration (i.e. k out of the n units need to function for the subsystem to function). Next we calculated the economic value of a 1 hour reduction of downtime for each unit in the system and used that value for the downtime penalty at unit level. The costs of a component failure would then consist of the costs due to do a repair (man hours and materials) together with the expected downtime multiplied by the downtime penalty.

Within a unit we had to consider more than 100 components which were addressed by a comparable number of maintenance activities, i.e. instrumental, mechanical or electrical. To keep the administration tractable and to utilize the advantage of maintaining similar components at the same time (in order to save set-up work) the activities were permanently grouped into maintenance packages. We did not try to optimize the contents of the packages. This was done using engineering judgement. Activities were also divided into safety related and non-safety related; for the first a fixed time schedule was used.

For the non-safety-related activities at component level we considered two basic models. The first was a mixture of an age and a block replacement model for revealed failures and the second was a simple inspection model for unrevealed failures. These models were developed in the 1960s (see Barlow and Proschan, 1965) and were well analysed. Yet they could not be used directly, since maintenance was restricted in our case to opportunities only. Also the few existing models on opportunity maintenance could not be used since these all assumed that opportunities were generated by failures of the components themselves, while in our case they were generated by causes outside the unit. Hence we developed new models (see Appendix A). From those models a priority criterion was constructed by considering the expected costs of deferring maintenance from one opportunity to the next one and subtracting from that the minimum average costs (see Dekker and Smeitink (1991, 1994) for more details). That is, at each opportunity we would calculate the expected costs due to deferring the execution of the package to the next opportunity and subtract from that the minimum average costs. Theory then yielded that deferring the execution of the package was cost effective (and thus smaller than the average costs) provided that the first opportunity was below the optimum threshold time for maintenance. The cost difference could then also be used as priority value in case multiple packages were due.

This idea applied to both the block and the inspection model, while the age model was approximated by an age-based variant of the block model (see Dekker and Roelvink, 1995; basically one uses the actual ages to calculate the cost of deferring the activity). During the development we only analysed the models roughly and used very simple approximations because of time pressure. For example, the renewal function (indicating the expected number of failures in a given interval) was initially approximated by the sum of the probability of one and two failures, and adding the probability of three failures from an exponential distribution with the same mean whenever it was assumed to be substantial. The assumption was that preventive maintenance would always be carried out before there was a substantial probability of failure. Only in a later stage were we able to finish the analysis of the models and derive better algorithms (see e.g., Smeitink and Dekker (1990) for a better approximation of the renewal function).

Three types of age measures were considered: run hours, starts and stops, and calendar time to relate the use of the machine to the probability of failure. Exponential filters were used to estimate on-line the average use in run hours and starts/stops, since these indicators had to be transformed to calendar time in order to allow predictions.

6.3.2 The furnace problem

The second problem was quite different, since a DSS was already available and the main task was to model the problem using this system and obtain the required data. Here one directly encountered a major problem. Since many maintenance problems are not repetitive, the DSS had been developed using generic problems and generic models. The problem at hand required a correct interpretation of the concepts used in the models of the DSS. In more detail, the DSS used concepts like failure and costs which had to be modelled by the user, yet in order to do so, the user required knowledge of what the model was doing with their interpretation. Although a local expert was available, the help of the company which developed the program had to be sought.

First the reasons to do preventive maintenance were identified: to bring the efficiency back to a better level and to prevent the possibility of failure. Since after preventive maintenance the unit was considered to be always in the same as-good-as-new condition the problem could be modelled with a mix of an efficiency model and an age replacement model (see Appendix B). All input was generated by expert judgement and most parameter values were quite uncertain (with an estimated relative error of 25 per cent).

6.3.3 The traditional engineering approach

Here we would like to contrast the operational research approach with the traditional engineering approach. The latter focuses on technology with much attention on failure analysis and a design out of failures. This has been a very successful route in many systems; compare, for example, the maintenance and reliability of a present-day car with one from the 1960s. It should be said that operational research does not yield better systems, only a better use of systems and application of scarce resources. Economics and operational research is still an underrated discipline in many engineering studies, and mechanical engineers are often blinded by technology. They also tend to consider all problems in a deterministic way, i.e. ignoring all kinds of random events. It is only recently that probabilistic methods have entered engineering.

6.4 Implementation

The operational DSS was developed on an IBM mainframe computer using a FOCUS data base and with several FORTRAN routines for optimization. It was tested extensively before taking it to the site. The actual data collection phase turned out to be a major task since maintenance management, with the new insights obtained in the initial study, decided to use a new definition of activity and package. After this engineering task was completed, a major difficulty was encountered with respect to reliability data at component level. The data present in the maintenance management system turned out to be unreliable and inadequate for the purpose. They were unreliable because personnel had been sloppy in data recording. (It should be said that different interpretations caused confusion; for example, in one case the actual downtime of a pump was registered, yet its magnitude turned out to be strongly related to logistics (spare parts) problems, rather than a true measure of the time required to repair the pump.) Furthermore, reliability data did not show many failures: for one-third of the failure modes there had been no occurrence. Finally, the data did not indicate the effect of maintenance (either a renewal or just an inspection repair – many activities involved inspection and cleaning). To overcome these problems we decided to make use of the judgement of the engineers to elicit statistical lifetime distributions. A questionnaire was set up and sent to the engineers at the platform (this had to be done by mail, since men worked in shifts and several were involved; sending an expert to the platform was considered to be costly). There was quite a difference in the answers between the experts. Sometimes the values were estimated relatively rather than absolutely, which is caused by anchoring (comparing two failure modes). This gave rise to extremely low estimates (once in a hundred years for a unique component). The questions had some redundancies: we asked for percentiles, two interval probabilities and a mean. Accordingly we encountered 'inconsistencies in the answers. There was little literature to help us in this respect, nor on elicitation (the available literature focused on probabilities rather than on distributions and, moreover, we needed two parameter distributions to allow for effective replacement). In the end we arrived at reasonable distributions, supported by the idea to update them regularly based on the reported experience. For more details, see Dekker (1989).

Both DSSs obviously used computing hardware to store the data, perform the analysis and report to the user. In the offshore case a maintenance management information (MMI) system was available onshore on an IBM mainframe. For the prototype DSS it was decided to make use of these existing facilities, given the large amount of data and the potential future integration with the MMI. Onshore staff ran the analysis and examined the list of advised maintenance activities which was then sent by internal mail to the platform. The offshore crew reported back on paper and data were fed manually into the computer. It was envisaged that in a later stage there would be a terminal on board the platform.

For the second problem things were different. Here the computer was used to evaluate a number of alternative values of parameters so as to obtain insight into what really drives the problem. Being able to quantify alternative solutions proved to be of crucial value. The decision had to be made in a team, with different players and different perspectives (production against maintenance). Without the ability to evaluate alternatives such discussions often tend to be yes–no games, with the more powerful party on the winning side.

6.5 Results

In this section we present some of the results of each DSS.

6.5.1 The operational DSS

Table 6.1 shows an example of a ranking list produced by the operational DSS. The user specifies the date at which an opportunity occurs and for which advice is required (this may be some date in the future). Next the user specifies the system, subsystem and finally the unit for which advice is requested. An important element in the decision procedure is the first alternative moment for maintenance, being the next opportunity. By default the system uses an historic three-point distribution for the time to the next opportunity, but this can be overruled. The table gives the ranking value (or priority criterion) for the first 12 maintenance packages addressing unit G-1070. It also shows the package code (indicating whether it is an M, mechanical, E, electrical, or I, instrumental package), a short description of

Table 6.1   Example of a ranking list produced by the operational DSS

```
               OPPORTUNITY MAINTENANCE ADVICE - NON-SAFETY MPs        *RLISTNS *
      INSTALLATION: CA   CORMORANT ALPHA      SYSTEM: A.06 MAIN OIL EXPORT
         SUB-SYSTEM: .05. PUMPING             UNIT: P-3060    AVON/MATHER&PLAT
   CURRENT OPPORTUNITY: 01 02 89
                                     -    0 DAYS LATER WITH PROB.   0 %
       NEXT OPPORTUNITY: 01 03 89 -  28 DAYS LATER WITH PROB. 100 %
                                     -    0 DAYS LATER WITH PROB.   0 %
   NO.  MP CODE   MP NAME           EFFORT    RANKING  EXECUTE
    1   M 23      VENTS               2.00       5.1   <——
    2   M 16      FUEL VALVES         1.00       -.0
    3   M 32      ANTI VIB PADS       3.00       -.1
    4   I 38      HI SPD SHUT OFF     8.00       -.2
    5   E 48      O/H MTR CONT CEN    8.00       -.2
    6   M 22      DAMPERS             3.00       -.3
    7   M 34      CELL BYPASS DOOR    4.00       -.3
    8   E 19      SWBOARD AUX SUPP    2.00       -.3
    9   E 21      AUX DIST BOARD      3.00       -.3
   10   E 41      O/H MTR CON CENT   16.00       -.4

   For further information enter MP NO.
   PF1      PF2          PF3      PF4      PF5     PF6       PF7     PF8     PF9    PF10
   SAFETY|             ||RETURN |       || NEXT  |PREVIOUS||       | HELP ||       |
        |             ||       |       ||SCREEN| SCREEN ||       |      ||       |
```

the package and the total man efforts required to execute it. The ranking value indicates the expected money lost by deferring execution of the package to the next opportunity, as specified previously. The user may get some more information (e.g. last execution dates) by pressing a specified button.

The advice was well accepted by the maintenance foreman responsible for the planning. It did not tell him what to do, it merely indicated the importance and time of each activity. He then had time to check whether the necessary spare parts were available and how many men with which skills (electrical, mechanical or instrumental) could be put on the job. At some time during the development of the DSS we considered building a knapsack module which would optimize the set of maintenance packages to be carried out during an opportunity. The idea was that the planner would enter the amount of time available and that the DSS would then give the best choice of packages (the required different man efforts to execute). Luckily we did not pursue this route; the ranking list reduced the problem to such an extent that the planner could make decisions in a short time span. The proposed knapsack procedure would take away the freedom to decide, whilst the procedure could not capture all aspects involved (stochastic durations, different skills available, unavailability of spare parts). Moreover, incorporation of a knapsack module would add extra complexity to the DSS and it would not save much compared with the human expert (in many model studies researchers compare 'their algorithm' with dumb procedures and never with the results from human experts; this is justified from a scientific point of view, yet in practice it is the human expert who needs to be improved upon).

After 12 months of operation an analysis was made of the benefits of the DSS. Clearly some 30 per cent less maintenance was being done and the backlog in preventive maintenance, till then inevitable, had virtually disappeared. On many occasions, the DSS correctly advised the offshore staff not to carry out preventive maintenance even if there was a good opportunity to do so. Quantification of true benefits could only be done, however, using the development model description of the process. Again the recording of activities in the MMI was too inaccurate to draw conclusions.

6.5.2 The strategic DSS

Here we will show the working of the strategic DSS on the furnace tube problem. First we specify the necessary data of the so-called base case, i.e. the case with the most likely values. We have to specify the deterioration in terms of an increase in energy and in terms of an increasing failure probability. After some discussion and fact finding the team arrived at the following data. Several characteristics had to be estimated. It should be said that one of the virtues of the DSS is that it forces people to consider various aspects, like costs and failure probabilities, which they would tend to neglect when the decision would be taken without use of the DSS.

6.5.2.1 Increase of energy use

- After 6 months: 5 per cent.
- After 12 months: 40 per cent.

These values are related to the energy use of a clean furnace: about Dfl 75 000 per month.

6.5.2.2 Lifetime data (according to a bathtub model)

A failure is defined as any event in which, by coking, the temperature rises to such an extent that either a leak occurs or the furnace has to be stopped to prevent such a leak occurring. In the bathtub model considered there are three phases. Maintenance-induced failures are assumed to occur only in the first phase; the second phase indicates a period in which there is no observable wear-out. This phenomenon is assumed to be apparent in the third phase. For phases 1 and 2 one has to specify the finish time and a survival probability. For the third phase one has to specify a survival probability for a free-to-choose interval. The values supplied to the program are:

- No maintenance-induced failures.

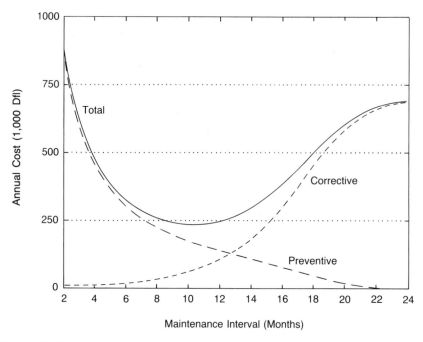

Figure 6.1 Long-term average costs.

- Chance of failure within 7 months after cleaning: 1 per cent.
- Chance of failure within 12 months after cleaning: 5 per cent.

6.5.2.3 *Maintenance data*

- Downtime required for cleaning: 7 days.
- Downtime required in case of failure, including repair and cleaning: 14 days.

6.5.2.4 *Cost data*

- Value of throughput: Dfl 600 000 per month.
- Direct cost in case of failure: Dfl 10 000 per occasion.
- Direct cost in case of a failure repair: Dfl 30 000 per occasion.

The DSS next applies a combined age replacement/efficiency model to this problem and produces a graph (see Figure 6.1) of the long-term average costs if the cleaning is done at regular intervals. In the graph a distinction is made

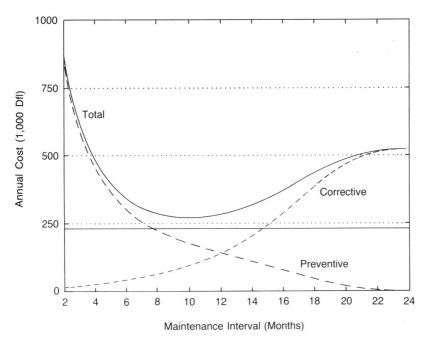

Figure 6.2 Average costs if the energy increase after six months is 10 per cent instead of 5 per cent.

between costs due to preventive maintenance (the preventive cleaning) and those due to corrective maintenance, in this case including the energy losses as well.

There is a considerable uncertainty about the values which have to be fed into the model. Consider, for example, the probability of failure 7 months after cleaning. If the furnace is unique, the data concern only one furnace and hence can hardly be based on historical experience. Besides, the latter would be appropriate for one time interval only! Accordingly, the failure probability has to be estimated and it could also be 2 per cent instead of 1 per cent. The DSS now supports this by doing sensitivity analyses, showing the effects of other input data. Figure 6.2 shows the average costs if the energy increase after six months is 10 per cent instead of 5 per cent. Note that the optimum interval hardly changes. Such an insensitivity is quite common in maintenance and helps to deal with the uncertainty in many quantities. After the team had done further sensitivity analyses, there was some confidence in the appropriateness of an 11-month-long maintenance interval. Next they redid the calculations with the cost of shutdown removed from the costs of preventive maintenance. It then appeared quite profitable to combine the cleaning with the shutdown of the processing unit.

6.6 Discussion

6.6.1 The value of the operational research approach

The operational DSS did make actual planning and scheduling a lot easier, by reducing the number of alternatives to be considered and by providing easy-to-handle priority criteria. It also had an effect on long-term strategy, rationalizing that maintenance which was not cost effective, allowing a better use of opportunities and hence reducing the workload on the annual shutdowns. It did enforce a coupling between machine importance and the amount of maintenance work spent on the machines. Although this is advocated in an approach like reliability-centred maintenance (RCM), it is not easy to achieve in practice, as RCM is only a qualitative approach.

A major disadvantage of the system developed was its need for data, requiring extensive data collections for some activities for which the optimization did not make a large difference. It would be better to develop a simpler procedure for some activities with less data requirements and less reporting effort, but which would also indicate whether a more advanced procedure was needed.

The advantage of the strategic DSS was in helping problem structuring. It did force people to think about economics and try to quantify various aspects which they were reluctant to do, because of a high uncertainty. Furthermore, the DSS did help people to work together rather than fight each other. In many cases there is a battle between the production and maintenance departments, with the former deciding when the latter may come into action.

One major advantage of a computerized DSS is the ability to evaluate the importance of data, just by evaluating the effect of different parameter values on the outcomes. This is of great help in getting a good insight into the problem. People often have a problem with costs which cannot be estimated exactly, like downtime costs, while only an order of magnitude is of importance. Another advantage of the DSS is the ability to evaluate many alternatives quickly in a rather objective way. This reduces the infighting between various people because each can make a proposal which is evaluated in an objective way. The outcome is of importance, but the other aspects may be more important.

A disadvantage of a strategic DSS is that it requires a central expert to operate it. Since each problem is different and appears only once in a while, expertise in operating such a system rapidly fades away. It is not always easy to model a problem with the models incorporated in such a system. There are many hidden assumptions and thoughtless use may even do more damage than good. Hence the DSSs should be very user-friendly and their models should be robust, flexible and transparent to the user. Development is not an easy task since it requires both expertise in the problem area (maintenance), in operational research models and in computer science. Moreover, there is a high obsolescence rate in computers, especially at the personal computer level. Computers and their software are old after just five years, yet within that time the developer should have sold enough of them to recoup the investment. Fortunately, new software development tools do make software building a lot easier and faster.

6.6.2 The literature of OR

In tackling two cases the gap between theory and practice became clear. The literature was not of much help in developing the described DSSs. Many models and algorithms had to be developed because existing methods were not appropriate for the problem. Applications require simple but effective methods. It did generate a lot of creativity, however, and even led to algorithms which can be published; see for example the simple idea for the renewal function (Smeitink and Dekker, 1990).

It is evident that in operations research there seem to be two different worlds. On the one hand there is the mathematical part which is preoccupied with solving defined problems in a nice mathematical way. This dominates the bulk of the scientific literature and in most cases the top journals. On the other hand there is the engineering area, with a lot of expertise on how practical problems should be tackled and how different disciplines, like statistics, operational research and computer science, can be merged. Little of this knowledge, however, appears in the literature. First of all it is more difficult to substantiate than the mathematical analysis of a simple model – establishing the optimality of the (s, S) rule is scientifically easier than defining the experience in reliability theory. Second, it is often not rewarding enough for practitioners to publish, whereas academics have to live by publishing research which is evaluated by their peers. Third, operational research academics often have no access to real-world problems, unlike physicists and chemists who can transfer their problems to a laboratory, as the real problems are in a non-scientific and rapidly changing industrial environment. Compare, for example, operational research with medical decision making: the latter is often housed within teaching hospitals, close to the problem owner, and the doctors, unlike managers, do have an interest in science. The medical decision literature is far more focused on linking theory with applications than operational research. Despite these bottlenecks for operational research we do think that people should try to bridge the gap, because models will only be appreciated in the long run if they are applied, and hence a discipline like operational research will only attract intelligent people if it has more than an artistic value.

Another dominating aspect of maintenance is the fact that it is often a learning process: there is not always enough duplication of units to allow the system reliability parameters to be learnt. Besides, failures are caused by many factors, many of which are influenced by local circumstances and use.

It is difficult to see where and how a system might fail – actual use is the final source of information, yet economic obsolescence is cause for early replacements. Finally, maintenance engineers are unfamiliar with economics and tend to maintain each unit in the same way, regardless of its importance.

Appendix A: the opportunity block replacement model

In the block replacement model a component is replaced preventively at fixed intervals of length t, which is the decision variable. Failure and successive replacement induces costs c_f, while a preventive replacement costs c_p. The lifetime of the component is a random variable with cdf $F(.)$, and pdf $f(.)$. Let $M(t)$ denote the renewal function corresponding to $F(.)$, indicating the expected number of failure replacements in an interval of length t. Elementary renewal analysis yields

$$M(t) = \sum_{k=1}^{\infty} F^{(k)}(t). \tag{A6.1}$$

For the long-run average costs $g(t)$ we have

$$g(t) = \frac{c_p + c_f M(t)}{t}. \tag{A6.2}$$

A typical example of the graph of the average costs is given in Figure 6.2. The extension of this model to opportunity maintenance is as follows. Suppose that the occurrence of opportunities can be described by a renewal process with interrenewal time Y and cdf $G(.)$. Let Z_t denote the excess time at time t, i.e. the waiting time to the first opportunity if t time units since an opportunity on which preventive maintenance has been done has passed. Z_t can easily be calculated if Y is Coxian-2 distributed (Dekker and Smeitink, 1991). In other cases, one can use the stationary excess distribution for which easy formulae exist. The average costs for the policy which maintains at the first opportunity at least t time units since the previous execution are given by

$$g_Y(t) = \frac{c_p + c_f \int_0^{\infty} M(t+x)\,dP(Z_t \le x)}{t + EZ_t}. \tag{A6.3}$$

Analysis of this model yields the following results. Let $\eta_y(t)$ denote the expected cost of deferring execution of a maintenance activity from the present opportunity at t time units since the past one to a next opportunity, Y time units away. Notice that

$$\eta_Y(t) = \int_0^{\infty} c_f[M(t+y) - M(t)]\,dP(Y \le y). \tag{A6.4}$$

We then have

$$\eta_Y(t) - g_Y(t)EY > 0 \Longleftrightarrow t > t_Y^* \tag{A6.5}$$

where t_Y^* indicates the optimal threshold for preventive maintenance and with equality if $t = t_Y^*$. Moreover, we also have

$$\eta_Y(t) - g_Y^* EY > 0 \Longleftrightarrow t > t_Y^* \tag{A6.6}$$

with equality only if $t = t_Y^*$. Notice that equation (A6.6) can be used as a priority criterion, since it expresses the expected costs of deferring the execution of maintenance to the next opportunity minus the minimum long-term average cost. The latter needs to be calculated only once and can then be stored in the data base. Evaluation of $\eta_Y(t)$ requires calculating only one integral numerically, which is no problem. Another advantage is that the meaning of the criterion allows a comparison with similar criteria developed for other block-type models, like the inspection model (see Dekker, 1995). This is an absolute necessity because priorities have to be set for many different activities.

Appendix B: a combination of the age replacement model and an efficiency model

In the age replacement model a component or system is replaced upon failure and when it reaches a critical age t. The age is set back upon each failure, contrary to block replacement where failures do not affect the replacement decision. Let $F(.)$ denote the cdf of the time to failure, c_p, c_f the costs of a preventive and a failure replacement respectively. Then the long-term average costs $g(t)$ of replacing at age t are given by

$$g(t) = \frac{c_p + (c_f - c_p)F(t)}{\int_0^t [1 - F(x)] \, dx}. \tag{B6.1}$$

In case the efficiency also drops in the course of time, we can extend the age replacement model in the following way. Let $E(t)$ denote the efficiency loss t time units after the last preventive maintenance, in percentages relative to the starting value, and let c_e denote the cost per time unit of a 100 per cent efficiency loss. We then have the following average costs when maintaining at age t:

$$g(t) = \frac{c_p + (c_f - c_p)F(t) + \int_0^t c_e E(x) \, dx}{\int_0^t [1 - F(x)] \, dx}. \tag{B6.2}$$

In general the integral in the denominator needs to be determined numerically, but this poses no problem. The theory of this model indicates the conditions under which there exists a finite age minimizing average costs and how that minimum can be found. Evaluating $g(t)$ for a number of interesting values of t often produces a good estimate of the minimum. This can be done with just a little computation time.

References

BARLOW, R. E. and PROSCHAN, F. (1965) *A mathematical theory of reliability*, New York: John Wiley.

BRITISH STANDARDS INSTITUTION (1984) BS 3811 – Glossary of maintenance terms in Terotechnology, London.

DEKKER, R. (1989) Use of expert judgment for maintenance optimization, Contribution to the first report of the ESRRDA Project Group on Expert Judgment.
    (1995) Integrating optimisation, priority setting, planning and combining of maintenance activities, *European Journal of Operational Research*, **82**, 225–40.

DEKKER, R. and ROELVINK, I. F. K. (1995) Marginal cost criteria for group replacement, *European Journal of Operational Research*, **84**, 467–80.

DEKKER, R. and SMEITINK, E. (1991) Opportunity-based block replacement: the single component case, *European Journal of Operational Research*, **53**, 46–63.
    (1994) Preventive maintenance at opportunities of restricted duration, *Naval Research Logistics*, **41**, 335–53.

SMEITINK, E. and DEKKER, R. (1990) A simple approximation to the renewal function, *IEEE Transactions on Reliability*, **39**, 71–5.

SOLOMON, L. (1989) Essential elements of maintenance improvement programs, Paper S4-7, IFAC Workshop on Production Control in the Process Industry, Osaka and Kariya, Japan.

VAN AKEN, J. E., SCHMIDT, A. C. G., VAN DER VET, R. P. and WOLTERS, W. K. (1984) A reliability-based method for the exploitation of maintenance opportunities, 8th Advanced Reliability Technology Symposium, B3/1/1-8, UK.

PART THREE

Computer-interactive OR

Single Component	Connected Subsystem	Complex System
Operational Decisions	Tactical Plans	Strategic Scenarios
Static	Evolutionary	Dynamic
Well-Defined	Multiple Concerns	Fuzzy
Small	Large	Huge

Fingerprints

- Problems typically deal with single components or connected subsystems, e.g. a vehicle dispatching problem linked to a larger planning system.
- The environment is rather operational or one where operations and tactics are difficult to separate.
- Though the environment may be static, e.g. 50 trucks to be scheduled every day, there are many small decisions to be made with a huge number of alternatives and therefore the problem often has a strong combinatorial flavour.
- Goals are fairly well defined though multiple concerns may have to be handled, e.g. balancing efficient fleet utilization with high service levels to clients with very strict time windows for delivery.
- The data set may be small or large but it is certainly very volatile and needs to be constantly updated.

This is where OR can support the experienced and knowledgeable operator who can no longer manage the decision and information overload. Given the often highly operational character of these applications, users are typically less educated and low in the organizational hierarchy. Though perhaps less versed in mathematics, these users do have much experience (*Fingerspitzengefühl*) which cannot easily be captured by algorithms. Computer-interactive systems present the users with alternatives. The latter remain in control, can add their own suggestions, and have the final say.

This is the world of user-friendly, visual, interactive planning systems. A pitfall may be concentration on the computer-interactive component at the expense of good, underlying solution techniques. Conversely, many applications fail because of insufficient attention to the user. Creatively balancing these two requirements is a challenge and makes this part a great playground for OR students.

The part contains four chapters with problems set in very different environments, ranging from shopfloor planning in parts manufacturing (Chapter 7) to scheduling of assembly lines in the fodder industry (Chapter 8), on the one hand, and from vehicle and driver scheduling in public transport (Chapter 9) to analysing an airline's staffing needs for maintenance engineers (Chapter 10), on the other.

CHAPTER SEVEN

Shopfloor planning and control for small-batch parts manufacturing

W. H. M. ZIJM

7.1 Introduction

This chapter describes the development of a planning and control system for small-batch parts manufacturing shops. A variety of case studies, almost exclusively performed at metal-cutting companies, motivated this development. The first case will be described in some detail to illustrate some key issues of the system developed. Other cases are briefly indicated with emphasis on the way they contributed to the basic framework and the (modular) set of planning and scheduling algorithms which constitute the algorithmic part of the shopfloor scheduling system *Jobplanner*, which is now a commercial package. Two objectives are pursued in this chapter:

1   to demonstrate how several, carefully selected case studies may lead to the development of a sophisticated OR-based tool, in this case a saleable shopfloor planning and scheduling system for parts manufacturing;

2   to show how *state-of-the-art theoretical knowledge may be used to develop methods* to solve complex real-world problems and, as such, to underline that the perception of a gap between theory and practice may be questionable.

To place this chapter in a proper context, a few general remarks on recent developments in manufacturing are in order. In particular, we indicate in what environments shopfloor planning and control systems are indispensable and when sophisticated operational research methods may add value in constructing such systems.

During the last three decades, dramatic market changes have caused many companies to fundamentally rethink their manufacturing strategy; cf. Hill (1994). Next to efficiency, quality and flexibility, both in production and engineering, also became key performance indicators; cf. Deming (1982), Blackburn (1991). In the context of this chapter, flexibility is defined as the ability to manufacture a large variety of products with short and reliable lead times. Basically, two ways can be identified in which companies have responded to these challenges:

1   to adapt manufacturing organizations to the changing market require-ments, often by changing from a departmental to a more product-focused manufacturing environment (e.g. cellular manufacturing, cf. Wemmerlov and Hyer (1989), but also business unit structures);
2   to install more sophisticated planning and control systems in an attempt to master the complexities of their manufacturing environment.

The above-mentioned changing market requirements can be characterized as a *market pull* impulse on changing production systems. Next to this, *technology push* instruments can be identified. The most dominant are without doubt the tremendous advancements in computer technology which have led to the development of computer numerically controlled (CNC) machines, flexible manufacturing systems (FMS) and automated materials-handling systems. Rapid advancements in software technology have further led to the rise of computer-aided design, computer-aided process planning and manufacturing planning and control systems. Finally, new technologies in manufacturing engineering should be mentioned, such as the use of new materials, miniaturization and rapid prototyping techniques.

A key question for many companies is how to exploit new developments in computer and information technology to adapt their manufacturing systems to changing market requirements. In particular the question whether existing hard- and software is suitable to fulfil the needs of a specific company is relevant here. To illustrate this point, let us focus on the application of MRP systems.

Computerized material requirements planning (MRP) systems were developed at the end of the 1960s to control the materials flow in large assembly plants; cf. Orlicky (1975). Starting with a master production schedule (MPS), basically a production plan for end items, the product structure (bill of materials) is exploited to calculate the requirements of subassemblies, parts and raw materials. Fixed offset lead times are reserved to enable the production or procurement of each subassembly or part. The absence of any capacity orientation led to the development of manufacturing resources planning (MRP II), a framework which adds various modules to support planning both at a longer-term production-planning level and on the shopfloor. A capacity requirements planning module may be used to check whether requested and available capacity match (cf. Vollmann *et al.*, 1992). Such a procedure works quite well in assembly systems or more generally in product-oriented production systems where a simple input–output control suffices, but it performs poorly in complex jobshop-type (functionally oriented) production systems, characterized by highly diverse routings of many different products. Karmarkar (1987) lists a number of reasons, e.g. the influence of actual workloads and of load-dependent lot sizes on actual lead times, to argue for a more sophisticated approach. Indeed, the influence of routing complexity is not adequately handled by capacity requirements planning modules.

Small-batch parts manufacturing shops are typically environments in which a high diversity of products for a variety of customers are produced in small lot sizes, which simply does not permit the installation of product-oriented fabrication systems. Typically, these shops need a shopfloor scheduling module between a higher-level global planning system (such as MRP II) and the actual machining system, to ensure that internally set *due dates* are met (in this chapter, due dates are defined as internal delivery dates for a parts manufacturing shop). Some companies even permit the alteration of *process plans* (a process plan specifies the technical data for producing a part, such as product routings, cutting tools needed and individual machine instructions), to relieve too tight capacity constraints (e.g. to reduce the load on a bottleneck machine). For such environments, the installation of a proper shopfloor planning and scheduling system is definitely in order.

Several shopfloor scheduling systems have been developed to fill the above need. Almost always, however, they primarily give administrative support, leaving it to the user actually to construct a schedule, after which automatic Gantt charts and a variety of reports are quickly produced. Some systems permit the application of straightforward priority rules, sometimes user defined, but the use of more sophisticated planning and scheduling techniques is rare (as is the case in MRP systems). One may of course wonder whether the use of sophisticated scheduling techniques is worth the effort. There are several reasons why we believe the answer to this question should be affirmative when posed in the case of modern, automated manufacturing systems.

One important reason is due to the inherent *multi-resource character* of modern manufacturing shops. The term *resource sharing* is often used to describe environments in which cutting tools, pallets, fixtures and operators are shared by various machines or even machine types. In more traditional fabrication shops, each machine typically possessed its own tool storage and clamping devices while operators were also often dedicated. In systems where only the set-up of products is performed manually, operator sharing becomes the rule rather than the exception, while the capital invested in high-precision tools and fixtures rules out the exclusive allocation of such auxiliary equipment to a single machine. Unfortunately, almost all scheduling systems are single-resource (machine) oriented, leading to a high probability that even a very short-term schedule cannot be maintained owing to the absence of auxiliary equipment. Clearly, priority rules do not help much since these rules are not able to handle the competition for resources between jobs adequately when these jobs need several resources simultaneously.

Even in traditional manufacturing shops priority rules may fail, in particular when product family structures appear. In such an environment, production of several products belonging to the same family can often be performed at the cost of minor changeover times; however, a major set-up time is incurred when changing to another product family. Hence, there is a natural tendency for machine operators to avoid set-ups by producing jobs belonging to the same family, thereby delaying jobs belonging to other families, resulting in a bad delivery performance for these latter jobs. It is precisely this phenomenon that we have observed in many machine shops, in which efficiency at one machine may cause idleness at another one (since the right job does not arrive in time), leading to an overall bad due date performance. As long as there are large amounts of work in process inventory on the shopfloor, one may be able to maintain a satisfactory degree of efficiency. However, demanding short lead times also leads to less workload on the shopfloor which immediately reveals the shortcomings of the traditional approach. Unfortunately, operational research also has long neglected these phenomena, where research was dedicated to either batching or scheduling but seldomly integrated both aspects (see, however, Potts and Van Wassenhove, 1992; Schutten and Leussink, in press).

Summarizing, we have argued that there is a definite need for sophisticated, OR-based, shopfloor planning and scheduling systems, in particular in modern, automated, small-batch parts manufacturing shops (a similar observation can be made for many semi-process manufacturing installations, although for partly different reasons). The rest of this chapter is devoted to the description of such a system, developed at the University of Twente. In Section 7.2 we present some background information on the initial case leading to this development. Sections 7.3 and 7.4 describe the basis of the methodology without giving algorithmic details (a list of references is provided for the technically interested reader). Results and contributions based on a variety of further cases are presented in Section 7.5. Finally, in Section 7.6, conclusions and some personal comments on key requirements when developing such a tool are given.

7.2 Manufacturing at El-O-Matic

The development of Jobplanner was highly motivated by a number of problems encountered at El-O-Matic, a medium-sized company producing pneumatic and electric actuators for valves (Slomp *et al.*, 1993; Slomp and Zijm, 1993). In this section, we start with a brief description of the basic characteristics of El-O-Matic, its market and products, as well as its aggregate planning procedure which marked the starting point for the development of the shopfloor planning system.

The firm El-O-Matic was founded in 1973 in Borne (The Netherlands) and started with the design and sales of pneumatic aluminium actuators. In 1981 the company moved to the city of Hengelo. In 1990, the company employed about 100 people in its main production facility in Hengelo, while small production facilities and/or sales offices were located in the UK, Germany, the United States and India. Starting with an annual sale of 11 000 pneumatic actuators in 1981, about 110 000 pneumatic and 10 000 electric actuators were sold in 1990. For 1995, a further increase to 177 000 pneumatic and 30 000 electric actuators was expected (of which 25 per cent was in the United States and 10 per cent in the Far East). This dramatic increase in sales volume (as well as product mix) forced El-O-Matic to rethink its manufacturing strategy as early as 1988.

In order to understand the change process the firm went through it is important to pay some attention to the products (the actuators) and the 'conventional' planning procedure. El-O-Matic distinguishes 12 basic types (sizes) of actuators. Each basic type can be delivered in a large number of variants. The most important parts of the actuator are the housing, two pistons, two end covers and the drive shaft. The diversity of actuators (the variants) is to a large extent determined by the actuator housings since these parts have to fit on the specific equipment of the customer.

The conventional planning procedure operated as follows. Each month, a sales forecast covering the next 13 weeks (one quarter) was presented at product family level. A product family corresponded to the basic actuator types mentioned above, while furthermore a distinction was made between products manufactured for the UK and for the Continent (we will return to this later), leading to 24 different product families. The company needed a total lead time of nine weeks, divided into three weeks for the procurement of raw materials (basic housings, shaft and covers, identical within one product family), two weeks for some basic parts manufacturing operations (variant independent), and four weeks for the variant-specific operations on the housings, some finishing operations and the assembly of the actuators (cf. Figure 7.1).

El-O-Matic promised customers a four-week delivery lead time (not including shipment and installation). The relatively low diversity during the first five weeks enabled the company to use forecasts to decide upon initial production quantities, only the last four weeks' production (consisting of variant-specific operations) was based on customer orders (we say that the *customer order decoupling point* (CODP) is located four weeks before actual delivery). It is clearly advantageous to reduce the variety during the forecast-based production to the utmost possible, to allow for more aggregate and hence more reliable forecasts; we will see that production automation may help to achieve this goal. Similarly, if there is pressure to reduce delivery lead times further, one should still seek for ways to perform all variant-specific operations during this reduced delivery lead time. Again, this may give rise to the need to automate specific parts manufacturing.

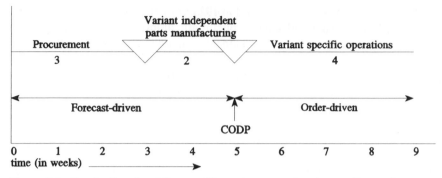

Figure 7.1 Production lead times with customer order decoupling point (CODP).

product family	total production (quarter)	time (weeks)												
		1	2	3	4	5	6	7	8	9	10	11	12	13
A	1200		300			300			300			300		
B	250	50		50			50			50			50		
C	450	90			90			90			90			90
D	400			100			100			100			100	
E	40		10				10			10			10
.	-	-	-			-		-		-			-	-
.	-							-						-

Figure 7.2 Master production schedule on product family level.

The planning department translated the aggregated forecasts into a basic production plan (actually a master production schedule on product family level), using only rough rules such as an economic batch size and an estimation of the required machine capacity (cf. Figure 7.2). The basic time period (time bucket) is one week. During the first five weeks of the total lead time, customers are invited to subscribe definitively to the scheduled quantities. These customers are, for instance, producers of valves and big oil companies. At the end of week 5, the subscription is frozen (or closed). If the planned quantities are not fully subscribed, some basic parts are kept in stock to fill any future demand (recall that the diversity is still relatively low after the first five weeks). Finally, the latter four weeks are used to complete the customer orders.

As mentioned earlier, the company was facing a rapid increase in both volume and mix in the late 1980s, resulting in a dramatic increase in required metal-cutting capacity. Since at the same time pressure on delivery lead times was experienced, the firm decided to install a flexible manufacturing cell (FMC) for performing a number of variant-specific processing steps on the actuator housings. This FMC and its associated planning and scheduling problems will be discussed in more detail in Section 7.4.

With the installation of the FMC some minor changes within the logistic planning structure were carried through but the main changes appeared to be a severe lead time reduction and a decrease of product variety before the CODP. The forecast-driven part of the production process still consists of the procurement of raw materials and only a few preprocessing steps (three weeks totally). For the latter production phases (i.e. all metal-cutting operations, including the FMC operations, followed by some finishing operations on the parts, and finally the assembly) three weeks are reserved (one week each), yielding a total lead time of six weeks. Since also the tapping of the screw thread is now performed in the FMC, together with all variant-specific operations, the number of basic housing types and therefore the number of product families has been reduced to 12. Hence, the diversity before the CODP declined to 50 per cent. During the latter stage, we may deal with more than 150 variants.

In order to reach the important lead time reduction during the variant-specific stage, and to exploit fully the potentials of the FMC, a sophisticated shopfloor planning system has been developed. This is the part of the story where operational research comes into play. Details will be given in the next sections.

7.3 A basic decomposition approach for hybrid jobshop scheduling

In Figure 7.3, a somewhat aggregated picture of the complete parts manufacturing shop of El-O-Matic is presented. The shop can be seen as built up from a number of work cells, where each cell itself may represent one processing step. The sawing machines perform some preprocessing steps (before the CODP). Other cells contain a number of parallel NC lathes, milling and drilling machines, a shaft processor, some standalone CNC machining centres, while the FMC also represents one cell. Each cell in itself may be quite complicated to plan, e.g. set-up times may be needed when changing from one product family to another, or additional tooling and fixturing constraints may play a role. For standalone CNC machines, set-up takes place at the processing platform of the machine, causing the machines to be idle. This is an important difference with CNC machines integrated in an FMC where set-up (building fixtures and clamping parts on a pallet) is performed off-line, after which the loaded pallets are automatically transported to the machines.

El-O-Matic uses an MRP system for global materials planning. Each week, a number of jobs are planned to be processed in the parts manufacturing shop. Each job may represent either a single workpiece or a very small batch of identical workpieces. Often, some jobs are still left from the preceding week (some overplanning is done regularly). Also, some jobs may be urgent. As a result, we are faced with a finite set of jobs, each with its own routing through the shop, while, furthermore, release dates r_i and due dates d_i for

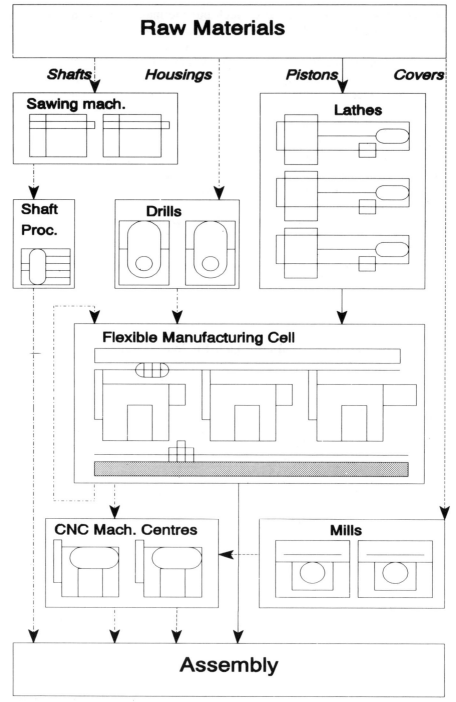

Figure 7.3 Parts manufacturing at El-O-Matic.

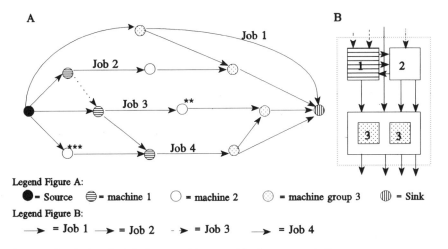

Legend Figure A:
● = Source   ⊜ = machine 1   ○ = machine 2   ⊙ = machine group 3   ⦀ = Sink

Legend Figure B:
⟶ = Job 1   ⟶ = Job 2   ⇢ = Job 3   ⟶ = Job 4

Figure 7.4   A simple jobshop and its graphical representation.

each job i are given. Let $c_i(S)$ denote the completion time of job i, given a schedule S. Then we seek a schedule that minimizes

$$\max_i\{c_i(S) - d_i\}$$

over all possible schedules S. The choice for this criterion function, i.e. minimizing the maximum lateness instead of minimizing the average lateness, was motivated by industrial experiences. Minimizing average lateness may lead to some very late and other rather early jobs, generally viewed as a less satisfactory solution, in particular when various part have to be combined in a subsequent assembly phase.

To understand the basics of the shopfloor planning and scheduling system developed, it is necessary to outline a fundamental decomposition approach, initially described by Adams *et al*. (1988). The algorithm of Adams *et al*. has been designed to minimize the makespan in a classical jobshop (one machine for each processing step) but is easily adapted to minimize the maximum lateness. Also, the extension to hybrid jobshops is possible, provided that good algorithms are developed for the single-stage scheduling problems which result after the decomposition (cf. Meester and Zijm (1994) for details).

The decomposition approach provides a mechanism to coordinate the schedules generated at the different work cells. Recall that at each work cell, only a single operation of each job is performed (this is not entirely true at the FMC but then different operations of the same job are treated as different jobs with a precedence relation between them). Suppose now that at the FMC each job is allocated to one of the three machines and that an ordering of jobs at each machine is given (we will say that jobs are *scheduled* on the

FMC although such allocations and partial sequences still do not determine starting times). However, these partial orderings of jobs on the FMC, together with the processing times of these operations and those of the preceding operations on other machines, easily determine the *earliest possible starting time* of the first operations after the FMC (given the initial shop release dates). Similarly, these partial orderings, together with all processing times of operations on the FMC and of any later operation, determine the *latest possible completion time* of the last operations before the FMC, to ensure that jobs are still finished in time (again, given the shop due dates).

A small example may help to clarify the procedure. In Figure 7.4, three work cells have to process four jobs; one work cell consists of two parallel machines whereas the other two cells represent single machines. Each node, except the source and the sink, denotes one operation (job–work cell combination). The solid arcs represent precedence relations between operations of the same job, and for each scheduled work cell the corresponding arcs are added. The length of an arc equals the processing time of the operation it starts from. Suppose that, one way or another, schedules for work cell 1 and work cell 3 have been determined (as indicated in the graph). Then, given the initial release and due dates of all jobs for the shop, a simple longest-path calculation can be used to determine the earliest possible starting times and latest possible completion times for the operations at work cell 2. These earliest possible starting times and latest possible completion times are called *virtual (operation-dependent) release and due dates of the jobs at work cell 2*.

The decomposition procedure now basically operates as follows. We start by calculating virtual release and due dates for all jobs on all machines, using only precedence relations of operations belonging to the same job (hence, routing information). Next, we exploit specific *work-cell-dependent algorithms* to determine a partial ordering of jobs at each work cell, and select the work cell that yields the largest maximum lateness. This work cell is termed a (temporary) *bottleneck*. We fix the job sequences on the bottleneck and recalculate virtual release and due dates of all operations on all other work cells (note that fixing a partial sequence adds arcs to the underlying graph and hence changes the longest paths). In this way we proceed by iterating through the work cells until all partial orderings are determined at all work cells and no further improvement occurs.

The above approach appeared to be very intuitive for operators and planners on the shopfloor in various manufacturing shops. The basic interpretation is that a partial schedule on one work cell induces the earliest possible starting times of succeeding operations and the latest possible completion times on preceding operations, given the initial release and due dates of all jobs for the shop. Note that such an approach results in an integrated scheduling solution, in contrast to myopic priority rules, which only consider the capacity needs of each job separately, but which do not

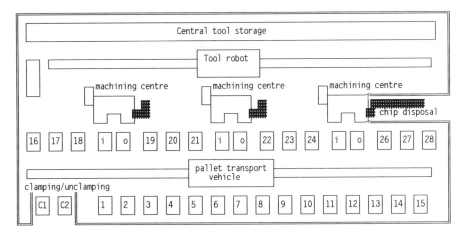

Figure 7.5   Schematic representation of the FMC.

handle the interaction of these jobs appropriately (e.g. in a schedule on one machine).

So far we have not discussed the solution of the individual (single-stage) work cell scheduling problems. As mentioned already, algorithms for these individual work cells depend highly on the structure of the work cell. These algorithms should attempt to minimize the maximum lateness of all operations at a particular work cell, with reference to the *virtual (operation-dependent) release and due dates* at that work cell. *Clearly, the problem is in the complexity of these single-stage scheduling problems.* A set of algorithms for different work cell structures has been developed in the course of the shopfloor planning and scheduling project. In the next section we will treat the scheduling of the FMC at El-O-Matic in some more detail; other algorithms, partly based on research performed at various companies, are briefly outlined in Section 7.5. However, note that the coordination mechanism, based on the decomposition approach, does not see any difference between these single-stage problems.

7.4   Planning and scheduling the FMC at El-O-Matic

Figure 7.5 gives a schematic representation of the particular FMC considered. Three basically identical machining centres (M1, M2 and M3) are linked together by a pallet transport vehicle and an integrated pallet buffer system, having a capacity of 28 pallets. Each of the three machines can hold a maximum of 40 tools in the tool magazine. The changing of tools in the tool magazines is performed automatically. A tool robot provides a connection between a central tool storage and the tool magazines. An FMC computer takes care of the coordination of all activities within the FMC. The clamping

Variant \ Tools	1	2	3	4	5	6	7	8
1	x		x		x			
2		x		x		x		
3			x		x		x	
4		x		x		x		
5	x		x			x		
6		x				x		x

Before

Variant \ Tools	1	5	3	7	2	6	4	8
1	x	x	x					
3		x	x	x				
5	x		x	x				
2					x	x	x	
4					x	x	x	
6					x	x		x

After

Figure 7.6 Part/cutting-tool table, before and after rearrangement.

and unclamping of parts on pallets are done manually on two integrated load/unload (L/U) stations.

The FMC at El-O-Matic uses a limited number of dedicated pallets and fixtures while many of the cutting tools are unique (no duplicates available), owing to the very heavy investment costs associated with these tools. On the other hand, the availability of the tool robot between the central tool storage and the local machine tool magazines provides additional flexibility since tools can be changed on-line, i.e. during processing on the machines. Nevertheless, a planner has to take care that no two jobs are processed simultaneously on two machines in the FMC if they need the same unique tool. In other words: the FMC planning problem is multi-resource in nature.

These problems are even more complicated because of the limited availability of pallet/fixture combinations. To avoid too many complications in the explanation, we will concentrate here on the planning and scheduling problem with machines and cutting-tool constraints.

The most common way to deal with tooling problems is to determine a loading solution for the machines (in which jobs and associated tools are assigned to individual machines within the FMC) before actually scheduling the jobs. Clustering techniques are often applied to group together those parts which are sufficiently similar with respect to their operational characteristics, resulting in a sufficient overlap in required tools. This becomes even more important if a large number of tools are unique (at El-O-Matic, 57 per cent of the cutting tools appeared to be unique). Basically, clustering boils down to the rearrangement of rows and corresponding columns in a variant/cutting-tool matrix, as illustrated in Figure 7.6.

Many algorithms have been developed to perform the clustering (with slightly varying objectives); see e.g. King and Nakornchai (1982), McCormick et al. (1972) or Kusiak (1986). In the case study performed at El-O-Matic, the algorithm of McCormick et al. performed best (see Gruteke, 1991). Note that in the case of El-O-Matic the capacity limits of the tool magazines do not play a major role, owing to the presence of the tool change robot. However, the clustering algorithm was implemented in a situation where the FMC was still logically (not physically) separated from the remaining part of the parts manufacturing shop, meaning that FMC operations constitute a separate planning level in the bill of materials. When integrating the FMC and the other parts of the shop in one manufacturing planning level (necessary to reduce further lead times), we actually need a *dynamic* clustering algorithm, taking into account the times when jobs become available for processing on the FMC, as well as their scheduled times for subsequent processing. To this end, a decomposition procedure for the FMC has been developed, similar to the one outlined in the preceding section, which boils down to iteratively scheduling both machines and unique cutting tools, where these schedules are tied together by the same virtual release/due date mechanism described earlier. A detailed description of the procedure is given in Meester and Zijm (1993); here we briefly outline the main ideas.

After decomposing the hybrid jobshop scheduling problem by the technique described in the preceding section, the problem left at the FMC is to schedule a set of jobs subject to virtual release and due dates (arising from the decomposition) on a set of basically identical machines with additional cutting-tool availability constraints. The main problem indeed concerns the coordination of the simultaneous availability of both machine capacity and unique tools. In the case where tools are not a real scheduling constraint the problem can be solved by applying a parallel machine scheduling algorithm (see e.g. Carlier, 1987; Zijm and Nelissen, 1990), after which any arising tool conflict can be solved heuristically (e.g. by a simple interchange procedure). However, when tools form a dominant constraint, it seems worth while to consider these tools separately. Basically, we sequence all jobs that use a particular unique cutting tool, then adjust virtual release and due dates of jobs on other tools, and next proceed with another tool. A somewhat simplified example may help to clarify the procedure. Consider Table 7.1. Only unique tools are considered. Release dates, processing times and due dates are denoted by r_i, p_i and d_i, respectively. G_i denotes the set of unique cutting tools needed to process job i, N_t the set of jobs that need the unique cutting tool t.

When considering cutting tools first, scheduling each set N_t on the corresponding tool t clearly leads to tool 3 as the first bottleneck tool. Note that some jobs using tool 3 also need tool 2; temporarily fixing the job sequence on tool 3 now induces adjusted virtual release and due dates for these jobs, thereby influencing the sequencing problem for tool 2. Again, we determine the next bottleneck tool and proceed by iterating through the cutting tools until a processing sequence for each individual tool is determined and no further improvement is possible. This procedure solves the so-called *tool exclusion* problem, i.e. it ensures that *no two jobs that need the same unique tool are scheduled simultaneously, by adjusting the virtual release and due dates accordingly*. Finally, we schedule all jobs, subject to the above adjusted virtual release and due dates, on the machines in the FMC, using the parallel machine scheduling heuristic of Zijm and Nelissen (1990). A detailed description of the complete algorithm, as well as a

Table 7.1   Six unique tools and 15 jobs, to be processed on the FMC

Job i	r_i	p_i	d_i	G_i
1	11	8	21	(3.5.6)
2	0	4	12	(1,4)
3	19	7	29	(1,2,4)
4	18	8	38	(2)
5	1	2	6	(2,5)
6	44	6	51	(6)
7	13	5	28	(1,3,5)
8	31	9	46	(4)
9	26	7	34	(3,5)
10	34	3	51	(6)
11	0	6	18	(2,3)
12	4	4	11	(1)
13	0	5	6	(5)
14	29	6	42	(4,6)
15	7	8	25	(2,3)

$N_1 = \{2,3,7,12\}$ $\rightarrow$ $S_1 = <2,12,7,3>$ $\rightarrow$ $L_{max} = -3$

$N_2 = \{3,4,5,11,15\}$ $\rightarrow$ $S_2 = <5,11,15,3,4>$ $\rightarrow$ $L_{max} = -3$

$N_3 = \{1,7,9,11,15\}$ $\rightarrow$ $S_3 = <11,15,1,7,9>$ $\rightarrow$ $L_{max} = 2$ $\rightarrow$ Bottleneck

$N_4 = \{2,3,8,14\}$ $\rightarrow$ $S_4 = <2,3,14,8>$ $\rightarrow$ $L_{max} = -2$

$N_5 = \{1,5,7,9,13\}$ $\rightarrow$ $S_5 = <13,5,1,7,9>$ $\rightarrow$ $L_{max} = 1$

$N_6 = \{1,6,10,14\}$ $\rightarrow$ $S_6 = <1,14,10,6>$ $\rightarrow$ $L_{max} = -1$

derivation of lower bounds for the optimal lateness and test results, can be found in Meester and Zijm (1993).

The reader may note that in the above procedure the cutting tools are basically treated as machines, or, better, as single resources which are optimized after decomposing the problem while the impact on all job parameters, and hence on other resources, is next evaluated. In that sense it is not necessary first to handle all the tools and postpone the parallel machine scheduling till the end. Basically, at each step of the iteration, one may verify whether a cutting tool or the machines constitute the bottleneck, and then focus on that bottleneck.

The above remark has a much broader implication. At any moment, any resource in a multi-resource scheduling problem may be considered separately, by adjusting the parameters of the jobs to be scheduled on this resource using longest-path calculations in the underlying graph. The difference between multi-resource scheduling and classical jobshop scheduling is the complexity of the underlying graph (nodes may have a degree higher than 2) but basically the decomposition technique remains unchanged. This is one of the main ideas behind the shopfloor scheduling system Jobplanner developed at the University of Twente.

7.5 Results and further extensions to a shopfloor planning and scheduling system

In this section we briefly discuss implementations and results arising from studies performed at El-O-Matic and a variety of other companies. We will not discuss numerical results of the algorithms developed; for these results we refer to the corresponding papers in the reference list. In general, one may say that the decomposition approach produces close to optimal results when applied to the classical jobshop, whereas the single-stage work cell algorithms either yield optimal results (e.g. when a branch and bound procedure is applied) or close to optimal results (e.g. the algorithm developed by Meester and Zijm (1993) for the FMC). Other algorithms are discussed in this section.

The combination of the procedures outlined in the preceding two sections was only partly implemented at El-O-Matic. This holds in particular for the clustering technique in Section 7.5 and a preliminary version of both the scheduling procedures in Sections 7.4 and 7.5. A reduction of the lead time from nine to six weeks as well as a reduction of 50 per cent of the number of basic actuator housing types occurred but it should be stressed that this is due to a combination of technological and planning/scheduling innovations. The main advantage of the integrated shopfloor planning and scheduling system is the possibility of integrating several formerly separated planning levels at a parts manufacturing shop in one planning level.

A further important practical result should be mentioned. A commonly heard objective against detailed scheduling is the occurrence of many disturbances on the shopfloor, making a detailed schedule of less use. Our observations at various companies revealed that many of these disturbances were *organizational* disturbances due to the absence of auxiliary equipment. Since the framework developed here explicitly considers this auxiliary equipment, the number of disturbances decreased substantially, leading to less nervousness and replanning in the shop.

Since the start of the project in late 1991, several studies in other companies have been performed, resulting in both the definition of additional work cell structures and algorithms to solve the related scheduling problems with virtual release and due dates. Altogether, this has resulted in the development of a complete multi-resource shopfloor scheduling system, called Jobplanner, in collaboration with a private company, consisting of former students of the Production and Operations Management Group. The system basically consists of three layers, an automatic scheduler, an interactive scheduling mode, and a monitoring and control system, while a central data base interfaces to the global planning system (often MRP) and the technical process planning system as well as to the shopfloor (the work cells) (cf. Figure 7.7). For further information the reader is referred to Heerma and Lok (1994).

In this section we concentrate on several extensions and experiences in

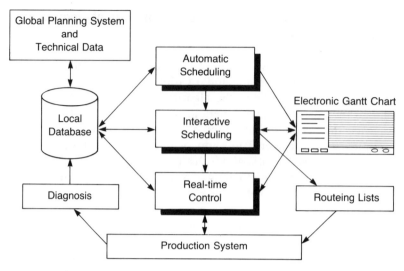

Figure 7.7   Architecture of Jobplanner.

other companies. Clearly, the main emphasis has been on the automatic scheduling functions although some interactive scheduling functions are briefly indicated.

Following Adams *et al.* (1988), work cells consisting of a single machine with jobs subject to virtual release and due dates without any further constraints are scheduled by Carlier's algorithm (Carlier, 1982). For parallel machines with release and due dates the heuristic of Zijm and Nelissen (1990) has been implemented. An important extension concerns the inclusion of so-called family set-up times. Here, jobs can be grouped in families according to similar set-up characteristics. On the one hand, one wishes to cluster jobs belonging to the same family, in this way saving set-up times, but on the other hand this may ruin the due date performance of other jobs which do not belong to this family (recall that all jobs are subject to due dates and that we wish to minimize the maximum lateness). This trade-off between efficiency and due date performance has been observed to be a basic problem in many companies. An efficient branch and bound algorithm has been developed by Schutten *et al.* (1996) and has been implemented in the system Jobplanner. A further extension to parallel machines with family-dependent set-up times is given in Schutten and Leussink (in press). Implementations of the technique have been performed at the Sheet Metal Factory of DAF Trucks in Eindhoven, The Netherlands. Test results revealed a considerable lead time reduction, in particular in the sheet metal press department.

Two case studies, performed at Morskate Actuators in Hengelo and at DAF Trucks in Eindhoven, stressed the need to schedule jointly the activities in the tool preparation room (calibration and assembly of tools with adapters, tool holders and the like) and the actual parts manufacturing shop. This extension is currently being investigated.

At a study performed at Stork Plastics Machinery, producing parts for injection moulding machines, it appeared necessary to include transport delays, not only to cover actual transport times but also to provide a certain robustness against small disturbances during execution of the jobs. Such delays are easily included in the decomposition procedure (basically the length of the arcs corresponding with the routings of the jobs are enlarged whereas all other arcs remain unaltered). Indeed this inclusion of delays provided the desired robustness without delaying the job completion times much (although sometimes the schedules changed significantly, when compared with the result of the procedure without transport delays). Note that when a transport facility constitutes a serious bottleneck (e.g. a limited number of automated guided vehicles (AGVs) or an overhead crane), it should be scheduled explicitly within the jobshop scheduling procedure, by considering it to be a separate workstation. Again, a bottleneck is defined here as the resource causing the largest lateness, *not* necessarily the most heavily loaded resource during a certain period of time.

A classical extension concerns the case of operators being responsible for multiple machines or machine groups. In such a case we have a joint machine operator scheduling problem (hence again multi-resource in nature). This happened to be the case in the machine shop of Ergon, a large electro-technical installation company in Apeldoorn, The Netherlands. Basically, a technique similar to the one described in Section 7.4 has been applied. The difference is that here unique resources (operators) have to be available at various work cells (which may subsequently be visited by the same job, as opposed to the situation described in Section 7.3 where the cutting-tool problem appeared *within* one work cell).

Next, we briefly mention a number of extensions that particularly apply to the interactive mode. First of all, a planner may wish to adjust the availability of machines (e.g. due to a breakdown or preventive maintenance activities, or even the absence of certain specially skilled operators). This calls for procedures for jobshop scheduling that are able to handle unequal availability of different machines in the shop (note that the classical jobshop scheduling problem assumes unlimited availability of each machine, or at least equal availability time windows). This problem can be handled by defining so-called dummy jobs with suitably chosen release and due dates. Algorithms which can be shown to determine optimally these release and due dates for dummy jobs have been developed. The same procedure applies if the operator wishes to fix certain jobs at certain completion dates. Details will be published in Meester *et al.* (1996).

A final extension which is currently being investigated is the integration of process planning and shopfloor scheduling, in particular for small-batch parts manufacturing shops in which process planning is often performed only a few days before actual job execution. In the process planning phase, job routings, required machines and cutting tools, and estimated processing times are often determined without taking into account the actual product mix on the shopfloor during execution (although this product mix can be rather safely predicted). Since the outcomes of the process planning phase form the input for the subsequent scheduling phase it seems worth while to consider the possibility of developing alternative process plans (e.g. in order to balance the load on the various machines in the shop during a certain period). This problem appeared to be crucial at Morskate Actuators in Hengelo. The critical path analysis provided by the Jobplanner system (and basically already present in the shifting bottleneck method) appears to be an excellent tool for detecting those jobs for which alternative process plans give the largest contribution to a further improvement of the due date performance. Details are presented in Zijm (1995); see also Lenderink (1994).

Alternative process plans as a tool to reduce lead times by better balancing the shops were also investigated at the machine shop of Urenco in Almelo, producing parts for ultra-centrifuges as well as for the environmental control systems for Fokker aircraft. The results were impressive. A lead time reduction of on average 30 per cent was obtained; cf. Wilbers (1994).

7.6 Epilogue

In this chapter we have described the development of a shopfloor planning and scheduling system for parts-manufacturing shops. The case of El-O-Matic, a producer of actuators for valves, has been described in some more detail to illustrate the application field of the system. Further contributions to the development of the system arose from experiences at various other plants. Indeed, the main purpose of this chapter is to show how a sequence of case studies may lead to the development of an OR-based decision support tool.

Most of the procedures mentioned above have been implemented in the current version of Jobplanner (cf. Meester and Zijm (1994) for an extensive review). The current version of the system costs about Dfl 30 000, when implemented as a standalone system. However, there is a major additional cost with regard to the definition of company-specific interfaces which connect Jobplanner to the global planning system, the technical process planning as well as to the shopfloor. Clearly, data analysis may further consume a considerable amount of time and money. Current development time of the system amounts to 4 man years of which roughly 50 per cent is devoted to the modelling and analysis of special work cell structures, and the design and tests of algorithms as described in Sections 7.4 and 7.5. The other 50 per cent primarily concerns software development.

In the following, we briefly comment on the role of OR in such case studies and in particular on what is needed to be successful in establishing satisfactory results. These comments are based on the experiences at El-O-Matic and the companies mentioned in Section 7.5, as well as on several other studies carried out at companies on different subjects. Most of the contacts with these companies are initiated as master's thesis projects for graduate students. In several cases, a long-term research contract evolves from these initial projects, depending on the initial results. A few key factors for success in these collaborations, as observed from these projects, are listed below.

Perhaps the most important key factor in projects with industry is a very thorough knowledge of the particular application domain. Note that the research described here has been performed within a *mechanical engineering department*, meaning that all our graduate students have a profound knowledge of concepts, models and methods in the field of *production engineering and production management*. In the case of El-O-Matic, the company would hardly accept a person having only a minor knowledge of manufacturing technology. Both students and researchers are primarily seen as experts on both the technical and the management issues of manufacturing. The fact that we apply OR-based tools in building, for example, design and control systems is taken for granted but the primary goal of the company is the effectiveness of the methods or tools we provide, not the underlying mathematical machinery. In most of the companies we have worked with, we discussed topics such as 'manufacturing systems design, planning and control, shopfloor scheduling' but never 'operational research', since the latter is viewed as a set of mathematical tools without any meaning as long as no interesting field of application is indicated. For the sake of clarity, the latter sentence does *not* express the personal opinion of the author but it certainly reflects the thinking of most of our counterparts in industry.

If the preceding key for success is accepted then the natural consequence is that one should concentrate on not too many different fields of applications. There are examples of successful OR applications in a wide variety of fields, ranging from financial areas to discrete manufacturing and process industries, logistics (physical distribution), strategic location analysis, planning and scheduling in service organizations, etc. But, as said earlier, a prerequisite of a consultant for acceptance is a good knowledge of the world he or she jumps in. This naturally limits the scope of one's activities.

It should also be emphasized that in all projects in which the author has been involved, model building and data analysis (including data modelling) have played an extremely important role. Good model building on the one hand requires that all *essential* elements of a physical situation or an organizational problem are captured, while on the other hand such a model should remain tractable, requiring a rather deep knowledge of existing solution techniques. Once having established model and data analysis, the development of algorithmic or heuristic solution methods was less time consuming. Their implementation, in particular programming and the development of software tools based on these methods, absorbed a major amount of time, although they required less intellectual activities.

The perception of a 'gap' between theory and practice does not reflect the author's personal opinion. Obviously, different groups of 'OR workers' can be identified. A substantial group of researchers in OR seems to be primarily interested in developing theories and methods. Such research definitely has its own merits; there is nothing more dangerous than saying that someone is working on a useless problem. Even if no direct application for the problem can be found, models and algorithms developed for it may prove to be valuable as a contribution to a sometimes unexpected other area. On the other hand, there are many people who primarily dedicate themselves to solving real-world problems with quantitative methods, which, if not more, is at least of ultimate importance for the development of the field. However, the perception of a 'gap' often reflects the opinion of practitioners that theoreticians develop models and methods which are not useful to them. In the author's opinion, the expectation of useful models and methods is not very realistic: OR offers a modelling and analysis *framework* (or call it philosophy) which is extremely useful when approaching problems at various levels within an organization. However, the author has never met any real-life problem which allowed the straightforward application of a well-known model or algorithm; at least a substantial modelling effort was required. We cannot blame the theoreticians for that; their task is to provide fundamental insights to problems which are often reduced to their most essential characteristics (completely similar to, for instance, fundamental research in physics), not to solve real-world problems (at least not in the first place). Hence, the relation between a good OR practitioner and theoretical OR should be similar to that between an engineer and theoretical physics. Unfortunately, experienced OR practitioners often show a lack of knowledge of the latest developments (e.g. control theory, queuing models in manufacturing), stating too easily that this knowledge would be useless anyhow. The ideal OR engineer should be able to apply the modelling and analysis approach, characteristic for OR, and should be aware of important recent developments in the field, always searching for useful elements but never blindly accepting them. In such a view, theory and engineering work together. Or, paraphrasing Niels Bohr, in OR there is nothing more practical than a sound theoretical model.

References

ADAMS, J., BALAS, E. and ZAWACK, D. (1988) The shifting bottleneck procedure for job shop scheduling, *Management Science*, **34**, 391–401.

BLACKBURN, J. D. (1991) *Time-based competition: the next battleground in American manufacturing*, Homewood, IL: Richard D. Irwin.

CARLIER, J. (1982) The one machine sequencing problem, *European Journal of Operational Research*, **11**, 42–7.

(1987) Scheduling jobs with release dates and tails on identical machines to minimize the makespan, *European Journal of Operations Research*, **29**, 298–306.

DEMING, W. E. (1982) Quality, productivity and competitive position, MIT Center for Advanced Engineering Study, Cambridge, MA.

GRUTEKE, R. (1991) Production control of a particular FMS, Master's Thesis, Laboratory of Production and Operations Management, Department of Mechanical Engineering, University of Twente (in Dutch).

HEERMA, W. and LOK, N. (1994) Scheduling systeem met IQ ontlast planner, *Logistiek Signaal*, **4**, 27–9 (in Dutch).

HILL, T. (1994) *Manufacturing Strategy: text and cases*, 2nd edn, Homewood, IL: Richard D. Irwin.

KARMARKAR, U. S. (1987) Lot sizes, lead times and in-process inventories, *Management Science*, **33**, 409–18.

KING, J. R. and NAKORNCHAI, V. (1982) Machine-component grouping in production flow analysis: an approach using a rank order clustering algorithm, *International Journal of Production Research*, **20**, 117–33.

KUSIAK, A. (1986) Application of operational research models and techniques in flexible manufacturing systems, *European Journal of Operational Research*, **24**, 336–45.

LENDERINK, A. (1994) The integration of process planning and machine loading in small batch part manufacturing, PhD Thesis, Department of Mechanical Engineering, University of Twente.

McCORMICK, W. T., SCHWEITZER, P. J. and WHITE, T. W. (1972) Problem decomposition and data reorganisation by a clustering technique, *Operations Research*, **20**, 993–1009.

MEESTER, G. J. (1996) Multi-resource shop floor scheduling, PhD Thesis, Department of Mechanical Engineering, University of Twente.

MEESTER, G. J. and ZIJM, W. H. M. (1993) Multi-resource scheduling of an FMC in discrete parts manufacturing, in M. Munir Ahmad and William G. Sullivan (eds) *Flexible Automation and Integrated Manufacturing*, pp. 360–70, Atlanta, GA: CRC Press.

MEESTER, G. J., WESTRA, S. and ZIJM, W. H. M. (1996) Job shop scheduling with unequal machine availability and fixed jobs (in press).

ORLICKY, J. A. (1975) *Material Requirements Planning*, New York: McGraw-Hill.

POTTS, C. N. and VAN WASSENHOVE, L. N. (1992) Integrating scheduling with batching and lot-sizing: a review of algorithms and complexity, *Journal of the Operational Research Society*, **43**, 395–406.

SCHUTTEN, J. M. and LEUSSINK, R. A. M. Parallel machine scheduling with release dates, due dates and family-dependent set-up times, *International Journal of Production Economics* (accepted for publication).

SCHUTTEN, J. M., VAN DE VELDE, S. L. and ZIJM, W. H. M. (1996) Scheduling a single machine with release and due dates and family-dependent set-up times, *Management Science* (accepted for publication).

SLOMP, J. and ZIJM, W. H. M. (1993) A manufacturing planning and control system for a flexible manufacturing system, *Robotics and Computer Integrated Manufacturing*, **10**, 109–14.

SLOMP, J., GAALMAN, G. and ZIJM, W. H. M. (1993) The impact of a flexible manufacturing system: a longitudinal case study, Working Paper, Department of Management Science, University of Groningen.

VOLLMANN, Th. E., BERRY, W. L. and WHYBARK, D. C. (1992) *Manufacturing Planning and Control Systems*, 3rd Edn, Homewood, IL: Richard D. Irwin.

WEMMERLOV, U. and HYER, N. L. (1989) Cellular manufacturing in the U.S.

industry: a survey of users, *International Journal of Production Research*, **27**, 1511–30.

WILBERS, A. A. B. (1994) Capaciteitsplanning by Urenco Nederland B.V., Master's Thesis, Laboratory of Production and Operations Management, Department of Mechanical Engineering, University of Twente (in Dutch).

ZIJM, W. H. M. (1995) Integration of process planning and scheduling in parts manufacturing shops, *Annals of the CIRP*, **44/1**, 429–32.

ZIJM, W. H. M. and NELISSEN, E. H. L. B. (1990) Scheduling a flexible machining centre, *Engineering Costs and Production Economics*, **19**, 249–58.

CHAPTER EIGHT

Production planning in the fodder industry

J. ROOSMA and G.D.H. CLAASSEN

8.1 Introduction

In the past decades Dutch animal husbandry has developed into an international branch of industry. It is without doubt that additional animal nutrients given by means of concentrate feeding have a significant influence on (farm) productivity. In order to cope with an increasing demand of concentrates and a continuous extension of the variety, the Dutch fodder industry developed into a limited number of large-scale production units. Figure 8.1 illustrates a simplification of a factory layout in the fodder industry.

In this branch of industry it is usual to buy raw materials in large amounts on the futures market. About 90 per cent of the raw materials are transported by ship to local production units where they are stored in large inventory cells. The size and recipe of a production lot determine how much and which raw material will be grounded and blended to meal. Usually the stock on hand of raw materials and the available capacity on the grinding and blending lines are amply sufficient. After the meal attains the right recipe, it will be pressed to pellets on so-called pressing lines. As this part of the manufacturing process requires steam pressure, the bits are collected in a cooler at the end of each pressing line. After a cooling period of approximately 30 minutes the final products are crumbled or directly conveyed to inventory cells from which trucks will be loaded.

During the cooling period no other products can be processed on the pressing line, which means that switching over from one product to another results in downtime (approximately 30 minutes). However, the available capacity of the pressing lines is limited. Consequently there is competition for the scarce resources between the different products. Additional difficulties are caused by dynamic demand, the availability of several non-

Raw material   Grinding and   Pressing lines   Inventory
    cells   blending lines   cells

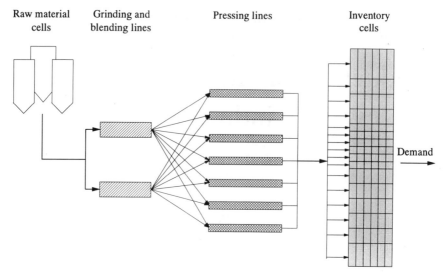

Figure 8.1 Simplification of factory layout in the fodder industry.

identical pressing lines and the needs or restrictions on the stockpiling of final products. In order to get a better understanding of the complexity of these kinds of problems we will focus in the next section on a real-world planning problem in the fodder industry.

8.2 The planning environment

The agricultural cooperative company 'ABC' annually produces well over 0.8 million tons of concentrates for about 5000 different customers in The Netherlands and Germany. In recent years many small production units have merged into a few large-scale factories located in the eastern part of The Netherlands. The largest production unit in Lochem is situated most favourably for the supply of raw materials by ship. For this reason especially most of the annual production takes place in Lochem. The product range consists of about 220 final products. The majority of this product range consists of slow-moving items which are not produced to stock. These products are strictly made to order, which means that their storage capacity equals zero. About 55 items can be held in stock for a number of periods in one or more fixed inventory cells. These so-called stock products encompass about 90 per cent of the annual production. The available storage capacity for final products is restricted to 900 tons for meal and 5500 tons for (crumbled) pellets. The meal storage is spread over 25 different inventory cells with an average capacity of 40 tons each. There are 128 different cells available for (crumbled) pellets and the content of these cells ranges from 8 to 100 tons. The production department is equipped with seven, non-identical, pressing lines. The potential capacity of each pressing line depends on the kind of product (i.e. bit size) that will be processed. Usually the available capacity of a line ranges from 10 to 30 tons per hour. However, as mentioned in the preceding section the losses of capacity are due considerably to downtime periods. In practice the maximum capacity of a pressing line ranges between 5 and 20 tons an hour. Production takes place continuously in three shifts from Sunday evening (24:00) to Saturday morning. The job arrival process can be classified as a deterministic dynamic shop with a planning horizon of five days. This means that new orders (jobs) are periodically released to the shopfloor (planning department) and the processing times of the intermittently arriving jobs are more or less known. Apart from the increasing number of last-minute orders, demand is known only one day ahead. Because of the limited quantity of actual customer orders in the orderbook, 90 per cent of the demand will be satisfied from inventory. In order to provide the required service level (backlogging is not allowed) demand is forecasted for the other days in the planning horizon.

In the current situation several planners in the production department are in charge of production planning for the next few days. This planning process is executed by hand and the principles of developing a plan are based strictly on experience. Daily information about the actual stock levels and (forecasted) demand for each item constitute the starting point of a production plan. Items for which the demand exceeds the actual stock level will get the highest priority. Products whose stock level will drop below the safety level will get priority next. As demand varies considerably within the planning horizon, a period-by-period planning approach will imply unbalanced stocks in other periods. As a consequence a planner tries to smooth the required capacity of the resources as much as possible. Moreover, the planner aims to produce large batches in order to avoid downtime on the pressing lines. Generally most of the (fast-moving) items will be processed on one or at most two pressing lines. However, there are only seven, not necessarily identical, lines available, which means that decision making not only concentrates on determining the size of the batches but also on the allocation of the planned lot sizes to pressing lines and inventory cells. Naturally, the restrictions of satisfying demand without backlogging and limited production and storage capacity have to be taken into account. Once the planning process has been completed, it turns out to be extremely difficult to cope with a sudden change in demand (rush orders) or breakdown somewhere in the production process.

The management team of the company had become aware that the current manual planning procedure was inadequate for coping with future planning problems. In fact, the motivation for developing a planning decision support system (DSS) was twofold. The first and most important reason was the extension of the production facilities in the factory; recently the number of processing lines has been extended from four to seven. Moreover, the storage capacity of the final products has been increased considerably. Secondly, the market is shifting from a bulk market to a market of special products. As a result the number of final products will increase in future. Without doubt this continuous increase of final products, together with a larger throughput, will lead to a more complex planning and scheduling process in future. Consequently the flexibility of the production department should receive a higher priority in order to meet the required service level, optimize the utilization of resources and minimize the set-up costs and inventory holding costs.

In the next section we describe how the problem has been handled and which difficulties were encountered in developing the DSS.

8.3 The OR approach

The problem is to determine a time-phased production schedule for multiple products on several, non-identical, parallel production lines, such that the sum of set-up costs and inventory holding costs is minimized. The time horizon consists of seven periods of one day on a rolling-horizon basis. Demand for each product has been specified per period and backlogging is not allowed. In the OR literature these kinds of problems are known as 'multi-item, multi-resource capacitated lot-sizing problems'. For the mathematical description of the problem we refer to model 1 in the appendix. Solving this mixed-integer linear programming (MILP) problem by an optimal solution algorithm would require a tremendous amount of computational effort because of the large number of (binary) variables. Therefore, efforts should be directed towards finding good and efficient heuristics for generating feasible solutions of high quality.

If we ignore the multi-resource aspects (seven pressing lines), the inventory capacity constraints as well as the losses of capacity caused by the changeover periods, the problem is identical to the well-known, multi-item, *single*-resource capacitated lot-sizing problem (CLSP). We refer to model 2 in the appendix for a mathematical description of this problem. Just like model 1, the CLSP belongs to a class of problems which are extremely difficult to solve within a reasonable amount of time. However, substantial research has concentrated on solving the CLSP with heuristics. According to Maes and Van Wassenhove (1988) the available heuristics can be subdivided into two classes: 'common-sense' heuristics and heuristics based on mathematical programming. The results as reported in the literature indicate that some 'common-sense' heuristics usually generate high-quality solutions with a relatively small amount of computational effort. However, these heuristics are unsuitable for solving our problem as formulated in model 1. The multi-resource aspects (several non-identical machines with different capacities), the inventory capacity constraints and the losses of capacity caused by downtime periods require several adjustments. Moreover, the planned lot sizes have to be allocated to a particular pressing line.

In order to reduce the complexity of model 1 we propose a solution procedure in two steps:

1   First an adjusted single-resource heuristic will be applied. This heuristic procedure takes the multi-resource aspects of the problem and the inventory capacity constraints into account, without completely altering the original concept of the single-resource heuristic. As a result the demand requirements can be clustered into lot sizes and scheduled on a fixed starting date.

2   The heuristic in the first step determines whether or not an item has to be processed in a particular period. This information is essential for solving the remaining problem. The problem here is to determine the actual size of the batches and to allocate them to particular pressing lines. As the preceding heuristic determines which of the items must be processed in each period, the remaining problem can be solved rather easily with model 1. Now, most of the (binary) set-up variables can be fixed in advance.

Many different 'common-sense' heuristics have been proposed in the literature for solving the multi-item CLSP. Basically, most heuristics consist of three steps: a *lot-sizing step*, a *feasibility routine* and an *improvement step*.

The lot-sizing step essentially consists of converting a given matrix of demands into a matrix of production lot sizes. By combining demands into production lots it is possible to save on set-up costs at the expense of some additional inventory holding costs. In order to decide on which demand quantities are included in a production lot, we applied the priority index as described by Dixon and Silver (1981). In fact this priority index is based on the well-known 'Silver–Meal' heuristic (Silver and Meal, 1973) for the multi-item, *uncapacitated* lot-sizing problem. The extension of the 'Silver–Meal' cost criterion by Dixon and Silver (1981) focuses on the competition for the scarce resources between the different items in the *capacitated* lot-sizing problem. Increasing the lot size of one specific item means a decrease in available capacity for the production of another item.

The second step in a single-resource heuristic is called the feasibility routine. This step ensures that all future demand can be satisfied without backlogging. Moreover, for some periods total demand can exceed total capacity and hence some inventory will have to be built up in earlier periods with slack capacity. Maes and Van Wassenhove (1986) distinguish two different approaches between the feasibility routines, i.e. a feedback and a look-ahead mechanism. Computational comparisons by these authors showed that the look-ahead procedure provides better solutions on average than a feedback routine. For this reason the feasibility routine in the forward-sweep approach, described by Dogramaci *et al.* in 1981, served as a starting point for some necessary adjustments (see Section 8.3).

Finally, the solution obtained so far can be enhanced by applying some additional improvement steps. Some researchers even included these steps as an integral part of their method (e.g. Dixon and Silver, 1981). We implemented only a particular improvement step called lot elimination (Dixon and Silver, 1981; Dogramaci et al., 1981). The effectiveness of an improvement step relies upon the possibility of demand splitting, whereas the lot-sizing step does not allow a demand to be split over more than one period. Of course a lot size will not be moved to earlier periods if this would violate the capacity constraints or increase the total costs.

As mentioned before, there are several heuristic procedures available in the literature for solving the single-resource CLSP. However, none of these methods can be applied directly to our problem. In fact, there are three major differences between our problem (model 1 in the appendix) and the single-resource CLSP as formulated in model 2:

1   The first difference concerns the inventory. For most products, the initial inventory will not be equal to zero. Furthermore, for all fast-moving items or so-called stock products the end-of-period inventory level has to be within certain bounds. The minimum level or safety stock is set by the planner while the maximum level will be fixed by the storage capacity of each product. In order to include the initial inventory level and to make sure that the safety stock requirements are met, we introduce the term 'net demand'. The matrices of net demands and preliminary end-of-period inventory levels are calculated before the lot-sizing step starts (Roosma and Claassen, 1994). Moreover, the lot-sizing step has been adjusted with a feasibility check in order to assure that the inventory capacity constraints will not be violated in any period.

2   Second, the single-resource heuristic does not take into account any set-up time. As our adjusted single-resource heuristic does not allocate the lot sizes to a particular pressing line, the downtime on each pressing line cannot be calculated exactly. As a result, an estimate of the daily available capacity has been used in the heuristic procedure. This inadequacy will be taken into account in the second step of the solution procedure (MILP).

3   The main difference, however, concerns the multi-resource aspects of model 1 versus the single-resource condition of most capacitated lot-size heuristics. In the case of multiple resources (e.g. pressing lines), the daily production capacity equals the sum of the production capacities of all the individual resources. The necessity for adjusting the feasibility routine would be almost superfluous when it is possible to process each item on any pressing line. However, since each item can be processed only on a restricted number of pressing lines, the feasibility routine has to be adjusted in such a way that the limited production capacity for the individual lines will be taken into account. The following example will illustrate the problem.

Table 8.1 Daily capacity per pressing line

Line	Capacity
1	363
2	242
3	121

Table 8.2 Incidence matrix for product/line combinations (1 = possible, 0 = impossible)

Product i	Line 1	Line 2	Line 3
1	1	1	0
2	1	0	1
3	0	1	0
4	1	0	0
5	1	1	0
6	0	0	1

Suppose there are three pressing lines available on which six products may be processed. The daily production capacity in tons per pressing line is given in Table 8.1. So, in each period t, the potential production is limited to 726 tons:

$$\sum_{i=1}^{6} X_{i,t} \leq 726 \quad \text{for all periods } t. \tag{8.1}$$

In fact, these are the kind of constraints which are usually included in a feasibility routine of any single-resource heuristic. However, as shown in Table 8.2, each item may be processed only on a few pressing lines.

Now we can see that the restrictions as formulated in (8.1) do not necessarily imply feasible solutions for our multi-resource problem. For example, suppose the equations in (8.1) are not violated; then it is still possible to plan a lot size for item 6 which exceeds 121 tons. This kind of infeasibility can be avoided by using a number of additional constraints (Roosma and Claassen, 1994). We adapted the feasibility routine of a single-resource CLSP heuristic (Dogramaci *et al.*, 1981) in such a way that these additional constraints will be fulfilled too. The necessary adjustments assure that the available capacity of individual pressing lines will not be violated by the planned lot sizes.

The idea in our approach is that the heuristic procedure in the first step determines whether or not an item has to be processed in a particular period. This should be done in such a way that the daily production and inventory capacity constraints are not violated and the set-up and inventory holding costs are minimized. The problem in the second step is to determine the actual size of the batches and to allocate them to particular pressing lines. As the preceding heuristic determines the items to be processed in each period, the remaining problem can be solved rather easily with model 1. Now most of the (binary) set-up variables can be fixed in advance.

The development of a prototype planning decision support system (DSS) started in 1992. From the beginning it was obvious that a regular and intensive dialogue with the planners of the company would be of crucial importance. In this way we gained both a thorough insight into the planning process and substantial support from the planning department. Special attention had to be paid to the development and implementation of a graphical user interface. The objective of the DSS is to support the planning activities concerning the items and the determination of the lot sizes to be produced on the different pressing lines. With a graphical representation of the solution the planner should be able to modify the computed plans in such a way that the solution will be tuned to the actual and future situation of the production department. After about one year a first release of the DSS was implemented on a powerful personal computer.

8.4   Validation of the system

In order to test the prototype planning system it had to be implemented and evaluated in a real-world environment. A software firm created an interface to the local mainframe on behalf of the daily data collection. For about two months the system has been run several times a day. Each time the generated production plan has been compared thoroughly with a solution drawn up manually by the planner. In most cases there occurred great similarity between both plans, which in turn led to great enthusiasm at the planning department. Moreover, the DSS made it possible to generate plans at any time and within a reasonable amount of computational time (a few minutes). In the current and future situation a planner will definitely need much more time to finish the daily planning problem without a DSS. Though the number of last-minute orders increases continually, a planner has to start the (manual) planning process before all the jobs are booked. This method implies that the remaining jobs will never fit optimally into the plan. Moreover, coping with rush orders is an extremely difficult task. With the help of a planning system a planner can postpone the planning task and generate plans at any time. The system has also been shown to be very powerful in generating alternatives or revised plans when unforeseen disturbances occur: for example, a breakdown somewhere in the production process or a sudden change in demand (rush orders). The potential value of the system can even grow when some time-consuming clerical actions and data processing will be carried out by the DSS.

However, as can be expected during a validation period, we also noticed some shortcomings in the system. With regard to the results, improvements should be made at the following points:

- support for the planner in case infeasibilities occur
- the processing sequence of the planned lot sizes
- the time scale of the planning horizon
- allocation of the planned lot sizes to inventory cells.

8.4.1 Support for the planner in case of infeasibilities

Occasionally an infeasibility occurred in the first step (the multi-resource CLSP heuristic) or in the second step (MILP) of the solution procedure. The existence of a feasible solution in the heuristic can be determined rather easily at the start of the lot-sizing step. Some help-screens have been developed to show the planner the reason for not finding a solution. Moreover, these screens offer some proposals for solving the problem. Sometimes the planners were faced with an infeasibility in the second step of the solution procedure (the MILP problem). As the multi-resource CLSP heuristic does not allocate the lot sizes to a particular pressing line, the daily decrease of capacity due to downtime, on each pressing line, is unknown in the first step. Apparently our estimation of the downtime seemed to be inadequate.

However, the test runs showed that this particular problem mainly occurred when several products did not fulfil the safety stock constraints in the first period. The results that were obtained after an extension of the production capacity in the first period confirmed our conjecture: the computed plans consist of many small lot sizes, especially in the first period. As a result the loss of production capacity, due to set-up times, in the first period was considerable. In order to avoid this kind of infeasibility we developed a simple heuristic rule. After this rule was implemented the generated plans turned out to be even more realistic and almost no infeasibilities occurred in the second step of the solution procedure.

8.4.2 The processing sequence of the planned lot sizes

So far we have only described a solution procedure for the multi-item, multi-resource CLSP. Moreover, the planned lot sizes were assigned to a particular pressing line. Actually the remaining problem can be formulated as determining the sequence in which the planned lot sizes should be processed on each pressing line. Note that the total costs in our objective function will not be influenced by the processing sequence since set-up time and costs are assumed to be independent of the sequence.

The sequence that will be chosen by a planner depends on several rules, such as:

- the lower the on-hand inventory level the higher the priority for a lot size;
- as contamination between products may cause serious problems, certain subsequent lot sizes should be excluded;
- transportation of different products from different pressing lines to the inventory cells at the same time can cause additional idle time;
- in order to meet the due dates of some orders a few lot sizes have to be processed before a fixed point of time.

The test phase showed that on this operational control level the human way of reasoning and experience is indispensable. In our opinion the (final) processing sequence of the planned lot sizes should be determined by the planner him- or herself. Of course, some computerized assistance (e.g. an interactive graphical user interface) could be beneficial.

8.4.3   The time scale of the planning horizon

A subdivision of the planning horizon into five planning periods, or days, turned out to be inadequate. The demand rate is time dependent and often specified per hour. Although 'long' planning periods permit multiple items to be produced in each period, an aggregation of the daily demand into a single period implies that a few orders will not be processed before their due dates. Moreover, a subdivision of the planning horizon into five days implies a lack of information during a period on the actual demand level, the production level and the inventory level for each product. However, a subdivision into shorter planning periods has a dramatic impact on the proposed solution procedure. At first the number of binary variables in model 1 will increase substantially. Second, one-day planning periods in this case study allow for multiple items to be produced per period and set-ups are prohibited to be carried over from one period to another, even if production of a given item occurs in successive periods. If the planning periods allow for at most one item to be produced per period, the set-ups have to be carried across period boundaries. As a result a modification of model 1 is inevitable. The losses of capacity caused by the downtime are no longer represented by one simple term in the capacity constraints. An adequate modification of model 1 means that the final processing sequence of the planned items must be taken into account. This in turn will have a considerable effect on the complexity of the problem.

8.4.4   Allocation of the planned lot sizes to inventory cells

As described before one-quarter of the products are stored in fixed inventory cells. In addition to the specified upper and lower bounds on the daily inventory level per item, the system should also allocate the planned lot sizes to the various inventory cells. Although model 1 can be adapted rather easily, an adequate adjustment of the multi-resource heuristic is far more complicated.

8.5   State of affairs

The aim of this project was to develop and evaluate a prototype decision support system (DSS) that generates and presents time-phased production schedules in a real-world environment, with a reasonable efficiency. It turned out that the proposed decomposition and solution techniques could solve a substantial part of the problem in a satisfactory way. The system has been shown to be very powerful in generating a feasible production plan, alternatives or revised plans within a few minutes. The target group showed great enthusiasm and appreciated the potential value of the prototype by its true merits. Meanwhile the management team has invited an OR consultancy firm to develop a final implementation.

Evaluating the pilot system in a real-world environment was of crucial importance. In this way we gained both a substantial involvement of the planning department and, as a consequence, a thorough insight into the planning process. Hence some essential shortcomings became self-evident. Partly, these deficiencies could be fixed during the test phase. However, the disadvantages as described in the Sections 8.4.3 and 8.4.4 have a considerable effect on problem complexity. In fact, it can be expected that the proposed solution procedure as described in Section 8.3 cannot be adjusted in an adequate way. The lack of flexibility in the second step, the MILP model, especially demands another approach.

Consequently the final implementation has been based completely on heuristics. Hence the system should not be considered as an optimizer but rather as a tool for generating production plans to be used for further analysis. In this connection the user-friendly and fully interactive user interface is of crucial importance.

Appendix: models 1 and 2

The multi-item, multi-resource capacitated lot sizing problem (model 1)

Mathematically the multi-item, multi-resource capacitated lot sizing problem of the company under consideration can be formulated as a mixed-integer linear programming (MILP) problem, in which the index i refers to the different items or (final) products ($i = 1, 2, \ldots, N$) and l to the pressing lines ($l = 1, 2, \ldots, L$). The index t denotes the specific day within the planning horizon T ($t = 1, 2, \ldots, T$). The following coefficients to be used in the model formulation are defined:

F_i the set-up costs for product i (in guilders)

h_i the inventory holding costs for product i per unit, per period (in guilders)

$d_{i,t}$ the given (forecasted) demand for product i in period t (in tons)

$IMAX_i$ the available storage capacity of product i (in tons)

$IMIN_i$ the desired minimum stock level of product i (in tons)

$CAP_{l,t}$ the available capacity of pressing line l in period t (in tons)

$k_{i,l}$ capacity absorption coefficient for product i on line l (in tons per unit)

a_l the loss of capacity at pressing line l, caused by switching over from one product to another (in tons).

Furthermore, the following decision variables are defined:

$X_{i,l,t}$ the volume of product i to be produced at pressing line l in period t (in tons)

$I_{i,t}$ the inventory level of product i at the end of period t (in tons)

$Y_{i,l,t}$ a binary (set-up) variable indicating whether or not product i is planned to be produced on pressing line l on day t.

Now, the problem under consideration can be stated as follows:

$$\text{Minimize}\left\{ \sum_{i=1}^{N} h_i \sum_{t=1}^{T} I_{i,t} + \sum_{i=1}^{N} F_i \sum_{l=1}^{L} \sum_{t=1}^{T} Y_{i,l,t} \right\} \tag{A8.1}$$

subject to

$$I_{i,t} = I_{i,t-1} - d_{i,t} + \sum_{l=1}^{L} X_{i,l,t} \quad \text{for all } i, t \tag{A8.2}$$

$$I_{i,t} \leq IMAX_i \quad \text{for all } i, t \tag{A8.3}$$

$$I_{i,t} \geq IMIN_i \quad \text{for all } i, t \tag{A8.4}$$

$$\sum_{i=1}^{N} (a_l Y_{i,l,t} + k_{i,l} X_{i,l,t}) \leq CAP_{l,t} \quad \text{for all } l, t \tag{A8.5}$$

$$X_{i,l,t} \geq 0 \quad \text{for all } i, l, t \tag{A8.6}$$

$$I_{i,t} \geq 0 \quad \text{for all } i, t \tag{A8.7}$$

$$Y_{i,l,t} = \begin{cases} 1 \text{ if } X_{i,l,t} > 0 \\ 0 \text{ if } X_{i,l,t} = 0 \end{cases} \quad \text{for all } i, l, t. \tag{A8.8}$$

Expression (A8.1), the object function, shows that the sum of inventory holding costs and set-up costs has to be minimized. The variable production costs are assumed to be constant for each product and equal for each pressing line. Therefore, the total production costs (excluding the set-up costs) will be constant and hence these costs need not be included in the model formulation. The constraints (A8.2) represent the production–inventory balance equations. The constraints (A8.3) ensure that the total inventory level for each product cannot exceed the storage capacity for that particular product ($IMAX_i$). The constraints (A8.4) express the minimum (safety) inventory level at the end of period t. These minimum stock levels ($IMIN_i$) are tools in hedging against uncertainties in future supply and future requirements. Consequently the set of equations in (A8.2), (A8.3) and (A8.4) assure that the demand $d_{i,t}$ for product i in period t can be fulfilled without backlogging.

The constraints (A8.5) represent the production capacity constraints: capacity absorption due to scheduled production and changeovers should not exceed available capacity. The constraints (A8.6), (A8.7) and (A8.8) set restrictions on the decision variables.

The single-resource capacitated lot size problem (model 2)

Mathematically, the single-resource capacitated lot size problem (CLSP) can be formulated as follows (Dixon and Silver, 1981):

$$\text{Minimize}\left\{ \sum_{i=1}^{N} \sum_{t=1}^{T} (h_i I_{i,t} + F_i Y_{i,t}) \right\} \tag{A8.9}$$

subject to

$$I_{i,t} = I_{i,t-1} + X_{i,t} - d_{i,t} \quad \text{for all } i, t \tag{A8.10}$$

$$\sum_{i=1}^{N} k_i X_{i,t} \le CAP_t \quad \text{for all } t \tag{A8.11}$$

$$X_{i,t}, I_{i,t} \ge 0 \quad \text{for all } i, t \tag{A8.12}$$

$$Y_{i,t} = \begin{cases} 1 \text{ if } X_{i,t} > 0 \\ 0 \text{ if } X_{i,t} = 0 \end{cases} \quad \text{for all } i, t \tag{A8.13}$$

$$I_{i,0} = I_{i,T} = 0 \quad \text{for all } i. \tag{A8.14}$$

In this model $X_{i,t}$ and $I_{i,t}$ represent the production and end-of-period inventory levels of product i in period t. $Y_{i,t}$ is a binary set-up variable indicating whether or not a set-up cost is incurred for the product i in period t. The parameters $d_{i,t}$, F_i, h_i, k_i and CAP_t are the demand, set-up cost, inventory holding cost, capacity absorption coefficient for product i and the machine capacity, respectively.

References

DIXON, P. S. and SILVER, E. A. (1981) A heuristic solution procedure for the multi-item, single-level, limited capacity, lotsizing problem, *Journal of Operations Management*, **2**, 23–40.

DOGRAMACI, A., PANAYIOTOPOULOS, J. C. and ADAM, N. R. (1981) The dynamic lotsizing problem for multiple items under limited capacity, *AIIE Transactions*, **13**, 294–303.

MAES, J. and VAN WASSENHOVE, L. N. (1986) A simple heuristic for the multi-item single-level capacitated lotsizing problem, *Operations Research Letters*, **4**, 265–73.

   (1988) Multi-item single-level capacitated dynamic lot-sizing heuristics: a general review, *Journal of the Operational Research Society*, **39**, 991–1004.

ROOSMA, J. and CLAASSEN, G. D. H. (1994) Production planning in fodder industry: a case study, Technical Note 94-05, Department of Mathematics, Wageningen Agricultural University.

SILVER, E. A. and MEAL, H. (1973) A heuristic for selecting lot-size quantities for the case of a deterministic time-varying demand rate and discrete opportunities for replenishment, *Production and Inventory Management*, **12**, 64–74.

Operational research methods and their application within the HOT II system of computer-aided planning for public transport

J. R. DADUNA and M. VÖLKER

9.1 Introduction

The fact that public transport companies are in a constant state of conflict is not a new one. They are permanently confronted with demands to produce an attractive, cost-effective service. At the same time they are forced to fulfil essential structural, political and social functions, the costs of which are not covered by revenue. The situation is comparable with squaring the circle – insoluble. However, there are ways and means of taking steps in the right direction, particularly with the use of computer-aided planning systems. A summary of system developments over the past years and their present state of progress are given in Wren (1981), Rousseau (1985), Daduna and Wren (1988), Desrochers and Rousseau (1992) and Daduna and Paixão (1995).

The HOT system is an example from this sector. It was developed in the 1970s by the Hamburger Hochbahn Aktiengesellschaft (HHA) to solve in-company planning problems in vehicle scheduling, duty scheduling and duty rostering. These problems show an extreme complexity so that to attain a feasible and also an efficient solution for such combinatorial problems, the use of OR methods is absolutely necessary. HHA for example operates a bus network within the Hamburg area, where about 2.4 million inhabitants live.

This bus network consists (in 1994) of 123 lines with a total length of 1482 kilometres. Up to 16 000 trips must be made per day. Approximately 10 600 trips are carried out by HHA, which require 780 buses and 2050 drivers, and 5400 trips by subcontractors.

The planning problem to be solved in the first step, the vehicle scheduling process, is to combine trips to blocks using an assignment algorithm. Each of these blocks can be carried out by exactly one bus, beginning with a pull-out and ending with a pull-in. In the next step, the duty scheduling process, the blocks must be modified, introducing additional restrictions (legal conditions, work regulations, in-company agreements). The duties which represent the workloads for the drivers are constructed on the basis of a heuristic procedure including a special version of an assignment algorithm with binary constraints. At the end of this planning process, the duties must be combined using an iterative assignment procedure to create duty rosters. This step leads to sequences of duties, which each have to be done by exactly one driver within a fixed period. For more details, especially concerning the mathematical models used in the HOT system, see for example Daduna and Mojsilovic (1988).

Since the application of the HOT system within HHA proved successful, other users have also adopted the system in the years that followed. In the late 1980s it had become necessary to redesign the system to conform with totally changed technical requirements, largely due to the explosive manner in which computer performance had been developing, particularly where the user interface was concerned (cf. Daduna and Mojsilovic, 1988). HanseCom, an HHA associate, has been responsible for this project for several years now (cf. Völker and Schütze, 1993).

The basic principle on which development of the HOT system was based was on how planning procedures could be simplified, results improved and the system utilized for simulation purposes. At the early stages of development OR methods were found to be of crucial importance. A drastic reduction in time-consuming manual procedures was aimed at, in order to create a wider opportunity for qualitative work at the same time.

The following sections summarize the present functions of the HOT II system and what experience and results have been gained from its practical application. As a system such as this is part and parcel of a comprehensive planning and operational control procedure, we shall begin with its incorporation and importance within the running of the company as a whole. We shall also include some basic remarks concerning the application of OR methods within a planning system in general.

9.2 Integration of the computer-aided system

It is essential that, when applied operationally, a system such as HOT II is not considered in isolation, even if it is fully functional from the operational point of view. What should be aimed at, rather, is a solution integrating

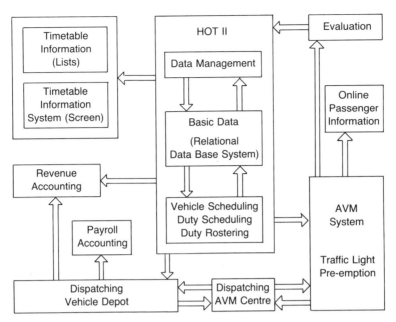

Figure 9.1   Integration of the HOT II system within the data processing structure at the planning and operational level.

planning and operational control, which is specially designed for the respective company and which comprises several different separate systems. There is no standard solution in this case – there never will be – but it must be designed especially for the individual size and structure of the company concerned (cf. Daduna, 1992; Petzold and Schütze, 1993).

Owing to its very conception, the system is an ideal foundation on which to work out an integrated solution such as this. The basis in this case is a data model comprising all network data required by planning and operational control. This information is kept in a relational database. Taking this structure as a starting-point, it is then possible to include all other systems which are relevant in this case via specially defined data interfaces. A detailed data structure such as this as a basis means that existing systems can be incorporated – an important fact from the companies' point of view. Seen graphically, the HOT II system (clearly set out in Figure 9.1) is a core surrounded by a shell of other preceding and successive systems.

Except where individual problems of strategic planning are concerned, the systems displayed in Figure 9.1 fully cover transport companies' requirements for the sectors of planning and operational control. If one studies the linkages between HOT II and the successive systems, it is evident that at this level only one-way relationships arise, i.e. direct data feedback does not occur. This does not mean that no allowance is made for using knowledge gained from the evaluation of operational running. This information finds a slot via the 'human interface', as revised data, particularly where service planning is concerned.

9.3 Outline of the HOT II system

HOT II is a modularly structured system which can be used in public transport companies of different size and structure, quite independently of whether public mass transport or regional transport is concerned. Nor does it pose limitations for sectors such as bus or tram. Neither is there a limit to the number of depots or to other companies being included which may take on work-load or pass it on. Legal provisions, work regulations and in-company agreements are included via parameters which can be modified if changes have to be made.

Modular structuring is based on planning functions within the transport companies. Each module comprises one phase of the planning process and is logically dependent on the previous one. The following are involved:

- Basic data management
- Sensitivity analysis
- Vehicle scheduling
- Duty scheduling
- Duty rostering.

Figure 9.2 demonstrates how these modules are connected structurally. It clearly shows that interactive and automated procedures – based on OR methods – significantly complement each other. This point is extremely significant, as these interrelations show that results obtained by automated procedures – using the system – are not final planning results but merely form a working basis for planning staff.

The system is designed for the user, which means it is used solely by planning department staff and not by staff working on data processing. For this reason system handling is by means of menus and specific screens. The menu structure depends on the operational sequence concerned but is not rigid. It can be structured for the user as both company and job specific. Screens and graphic functions are available for data handling and for operating purposes.

A description of the different modules is given in the following sections. On the whole, presentation is confined to functional purposes, with the emphasis on those parts which have direct connections to OR methods.

9.3.1 Basic data management

The data structures of the HOT II system cover all requirements arising within a mass transport company's planning and operational control sectors.

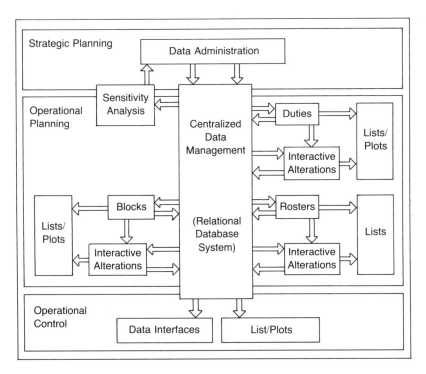

Figure 9.2   Structure of the HOT II system.

By using central data administration for operationally relevant data backed up by the ORACLE database system, efficient data management can be guaranteed where performance and consistency are concerned.

A relational database system such as this fulfils two basic principles intrinsic to essential efficiency. Each piece of information is only stored once, which ensures that no data inconsistencies occur owing to incomplete updating. In addition, because of the detailed, flexible data structures involved, it is possible via the data base system to edit information required for supplying successive systems in a manner specific to the needs of the subject company.

The basic structure of the data model used is a network one. It is made up of nodes and arcs (route sections), which can be specified comprehensively by numerous different attributes. This manner of representation enables existing data to be expanded step by step at any time, whether network information is increased or additional network areas are included.

The data set is divided into timetable periods, which, in their turn, are subdivided into day types MF (= Monday to Friday), SA (= Saturday) and SO (= Sunday). Structuring of this kind is necessary so that different planning phases can be handled parallel to one another on a common set of basic data.

9.3.2 Sensitivity analysis

The sensitivity analysis module is only found in this form in HOT II (cf. Daduna, 1988; Daduna *et al.*, 1993). The underlying principles for its development were considerations as to how aspects of operational performance could be included when timetables were compiled, so that greater efficiency would be obtained in operations. As simultaneous planning had to be ruled out owing to the extremely complex nature of a planning problem such as this, sensitivity analysis presented an opportunity which led to existing timetable data being reviewed from the vehicle scheduling angle. The scheduling problem becomes modified using a relaxation of timetable data to raise the degree of freedom in the combinatorial process.

Sensitivity analysis is particularly applicable in areas considered critical from the operations point of view, i.e. at rush hours. Peak-load periods generate vast cost, because the maximum number of blocks during these periods dictates vehicle requirements and therefore also indirectly the number of duties required. This inevitably leads to one fundamental objective: a reduction in the number of blocks during rush hours, not only by using optimization methods to work out vehicle scheduling, as is usually the case where fleet dispatching is concerned, but also, as in this case, by making changes in timetable data. Changing timetable data does not mean that travel times or layovers between the end of one trip and the beginning of the next – fixed by work regulations or stipulated by in-company agreements – are then freely available. Only departure times and buffer times may be changed, which are set down by the planning staff. In other words, general in-company restrictions and existent service levels remain unchanged.

Sensitivity analysis works in three steps: data editing, which includes establishing a start solution, interactive operating procedure and updating of the data base.

1   *Data editing and establishing a start solution (setting up a model).* The user must first define which day type and which interval of time he or she wishes to process and which maximal alterations are considered admissible. This information is used as a basis for building up the necessary internal data structure and, via vehicle scheduling (see also Section 9.3.3), a start solution which displays the number of blocks within a given period of time, without making alterations.

2   *Interactive processing (application of algorithms).* Once a start solution has been found, various suggestions for improving the existing solution are then called up by means of a modified assignment algorithm. The user checks these suggestions related to operations and either accepts or rejects them. The procedure can be terminated by the user or it ends when the system has no more suggestions to make.

3   *Updating the database.* Accepted alterations are not stored immediately

in the database. Any data changes are printed out first, to enable detailed checks to be made.

Sensitivity analysis is essential for implementing efficient planning but in no way can it replace the human element. Any alterations recommended by the system are always checked by the user for possible consequences within the operations and for their possible influence on the quality of service level offered (schedule synchronization etc.).

As parameterized controlling occurs when sensitivity analysis is used, it is of course possible to make a rerun at any time using different parameters. Reruns can also be made to modify individual alterations within the context of the operational procedure.

9.3.3 Vehicle scheduling

Vehicle scheduling is of particular importance when considered under the heading of 'resource scheduling', particularly where subsequent duty scheduling is concerned. At this stage results produced via sensitivity analysis are improved by reducing non-productive periods between trips on service. The number of blocks during peak periods remains unchanged. The principle here is to apply an efficient assignment algorithm.

Vehicle scheduling is worked out separately for each day type, taking every trip on the timetable into consideration. It is also possible to select subsets of trips, and a variety of different in-company regulations can also be included. One can even allocate specific lines or single trips to one or more depots.

Vehicle scheduling is divided into three steps which are carried out one after the other: preparations phase, automated scheduling and interactive alteration.

1 *Preparations phase (setting up a model)*. If it makes sense for operations, specific trips can be linked with each other permanently to define blocks or part blocks. For the MF day type trips are then merged which vary depending on the day of the week, so that a uniform structure is brought about for planning purposes. Such 'representative' trips are then added to compensate for any deviations. These structures are reset again once duty scheduling is complete (see also Section 9.3.4).

2 *Automated scheduling (application of algorithms)*. Automated scheduling first involves setting up a model by creating a matrix which contains all admissible links between two trips within the given in-company restrictions. Using an assignment algorithm, trip sequences are then constructed from this matrix. At the next stage these sequences are distributed between depots, if several exist. This is where a transportation algorithm is used. The criterion for the allocation of these sequences is the length

of pull-out and pull-in, whereby the feasibility of an allocation and the vehicle capacity available at different depots are taken into consideration. The result of this automated processing is a complete vehicle schedule, in which all existing restrictions are taken into account. This schedule is primarily a working basis for the user, whose job it is to check in detail when and at what different points it is possible to make manual improvements within the existing in-company restrictions.

3 *Interactive alterations.* Both graphic and alphanumeric functions are available for schedule alterations. These can be used independently from or combined with each other. The user has a choice here, which enables him or her to adapt the working method to his or her own individual concept. Apart from schedule alterations, individual trips can also be added or altered. Once accepted by the user, all alterations are stored immediately in the database.

9.3.4 Duty scheduling

Duty scheduling, whereby individual duties are established for drivers, is based on the results obtained from vehicle scheduling. This next stage of planning is the transition from solving a vehicle-orientated problem by including additional parameters – ones needed to define the work of the driver.

Duty schedules are usually established separately for each depot and according to day type. This does not mean, however, that relocation of certain duties to another depot is not possible within interactive alteration procedures. It may even mean, in a modified version of HOT II, that two depots are dealt with simultaneously. An important objective of the duty scheduling process is to adhere to the number of vehicles required during peak hours which was already worked out at the vehicle scheduling phase. Also unproductive time periods, in so far as these are not fixed by legal conditions, work regulations or in-company agreements, are removed from schedules as far as possible.

Here too, on analogy with vehicle scheduling, three steps arise: a preparations phase, automated duty scheduling and interactive alterations of the duties worked out by the system:

1 *Preparations phase (setting up a model).* Here it is possible for the user to fix specific duties and/or duty segments. These may, for example, be duties for specific groups of drivers (maintenance staff, works council members, part-time drivers etc.). By fixing these details a certain section is taken out of the process and removed from further duty scheduling, only then appearing in the duty schedule printout.

2 *Automated duty scheduling (application of algorithms).* Automated duty scheduling has to include a whole number of restrictions such as maximum

length of duty, maximum spread time, maximum driving time and/or break regulations. Where break regulations are concerned the system will accept different structures. Either a paid proportional break can be worked through, whereby the proportion is variable, or the regulation based on paid block breaks applies (1×30 minutes, 2×20 minutes, 3×15 minutes). The alternative is a module which enables unpaid block breaks of varying lengths to be included in the calculations. Taking as a basis the given blocks, duty segments are arranged, whereby a set of rules has to be adhered to. For example, a schedule may not be cut during peak hours as this would engender additional vehicles and drivers. Reliefs should also be avoided at these times, as these lead to an increase in driver requirements. As long as they do not already form a complete duty, the duty segments established at this stage of the procedure are then combined to form complete duties by an automated process supported by OR methods. If the block break regulation applies, all duty segments are included in the scheduling process. As a result feasible duty schedules are produced which satisfy all given restrictions.

3   *Interactive alterations.* Interactive alterations for duties are carried out by using alphanumeric functions.

As aggregate structures are used when scheduling MF day types (see Section 9.3.3), the final step of driver scheduling is to produce differentiated duties for each day of the MF type. This duty schedule can either be printed out as a graphic plot or as a list.

9.3.5 Duty rostering

During duty rostering, duties are combined to create rosters. These form the basic structure for planning driver assignments. Processing in this case is carried out by incorporating all day types separately for each depot. The objective is to work out well-balanced rosters which adhere to all necessary work regulations and in-company agreements, so that the resultant workload is distributed as evenly as possible amongst the staff and additional staff expenses due to overtime are avoided. Different rosters types are available for this procedure.

On analogy with the other modules of computer-aided scheduling, three steps also arise here: a preparations phase, automated duty rostering and interactive alterations of the rosters built up by the system.

1   *Preparations phase (setting up a model).* Here the user is primarily given the chance to make changes in the schedule structure. In this case specific duties are removed from the automated procedure and/or allocated to specific groups, in so far as different rosters types are used.

2   *Automated duty rostering (application of algorithms).* Taking the quantity

of duties available as a basis, rosters are then worked out with the help of an assignment algorithm, using a multi-level procedure. Here it is important not only that the legally stipulated overnight rest period be adhered to between the end of one duty period and the beginning of the next, but also that basic rosters structures be followed in accordance with in-company agreements (arranging duties according to duty type).

3   *Interactive alterations.* Interactive processing of duty rosters is carried out by using alphanumeric functions.

Duty rostering forms the final step in operational planning. The rosters and the data established at the foregoing planning stages form the schedules to be used as a basis for operations. It is therefore clear that, where a computer-aided system is used, the efficiency of operations will depend considerably on the efficiency of the system.

9.4   Use of OR procedures

The different possible applications of OR methods within a computer-aided planning system such as HOT II are always a constant source of controversy. On the one hand they provoke total dismissal, partially on the grounds that users of computer-aided planning systems are against automatically generated solutions. On the other hand they widen the scope of expectation, which is, after all, based on manual planning procedures being transferred to automated operations and in which the human element plays only an indirect role. Unbiased discussions have proved that neither of these arguments is tenable.

If proper assessment of these systems and their functional efficiency is to be made, certain basic points have to be considered:

- First, careful attention should be paid to the fact that, in a model, reality is abstracted. This step is necessary because, owing to their complexity, the given structures do not make allowance for OR methods being directly applied, i.e. in the respective model there is necessarily a limitation to problem-specific variables.

- On analogy, this also applies when formulating objectives. In reality a system of objectives will produce interdependencies and contradictions which result from efforts to make allowance for several objectives at once. This means that at this point, too, a functionable objective has to be set, which *per se* involves compromises.

- Setting up a model and formulating an objective lead to a formalized problem, but one which can be solved by using the appropriate OR algorithm. The solution, once found, will then only apply to the formalized problem and will have to be projected back into reality in an additional step. If optimization procedures are used then optimality will, of course, only refer to the formalized problem and not to reality.

This procedure forms the foundation for the efficient application of OR methods in planning systems. After all, only if all this is taken into account during the development and application of systems of this kind can they find acceptance by the user. The latter is not interested in which of the available OR methods is being used. Most important for the user, apart from the quality of results, is how the user interface is designed, which includes how results are displayed and what possible interactions can be taken when setting up a model.

The philosophy behind the HOT II system is based on these fundamental considerations. It works as a dual process for each planning phase (apart from basic data management) in which OR methods and procedures for interactive alterations of the solutions built up by the system form a balanced whole.

The user basically has two phases in which to operate: the preparations phase for automated procedures and interactive alterations to the schedules built up by the system (see Section 9.3). The user also has an indirect influence on automated procedures, whereby in-house experience gained in the user's own organization is linked in via the relevant parameters when model structures are being set up. This aspect was particularly important when the HOT system was being developed, which was originally conceived for in-company application. This meant that, from the very beginning, there was close cooperation with the planning staff.

It was only possible to attain the experience and results described in the next section by working in this manner. It would never be possible to obtain comparable solutions by using only a computer-assisted manual procedure or only automated techniques. This applies both to operational aspects and cost structures.

9.5 Experience and results

Practical use of the HOT system in public transport has already been made for over 15 years now. At first it was brought into operation for HHA, but as mentioned above it was successively implemented at other public transport companies in Germany as well as in France. The Kölner Verkehrs-Betriebe Aktiengesellschaft, which uses the system for planning bus and also tram operations, started at the beginning of the 1980s. Step by step other companies followed, as for example the Gesellschaft für Straßenbahnen im Saartal (Saarbrücken), the Stadtwerke Trier, the Verkehrsbetriebe Hamburg-Holstein Aktiengesellschaft, the Pinneberger Verkehrsgesellschaft mbH, the Bremer Straßenbahn AG, the Compagnie des Transport de Besançon and the Koblenzer Elektrizitäts- und Verkehrs-AG. Besides these, other public transport companies are starting to implement the HOT system. For some of these implementations the system has been extended by company-specific modules, especially concerning the duty scheduling procedure and the design of different printouts.

Experience gathered over this long period played a decisive part in the further development of the system. Apart from constant improvement of the algorithms applied the system was adapted to changes in requirements demanded by the user, particularly where data integration and consistency, design of user interface and improved control by the use of parameters were concerned. The points listed below refer to the most recent experience and result primarily from the system functions available today.

The quantitative improvements undoubtedly produced by successful in-company introduction of the HOT system at the planning level cannot on the whole be reconstructed explicitly. One reason for this is the complex structure of the problems, others are based on concrete facts:

- Transport companies' planning staff are under permanent pressure, particularly when vehicle and duty schedules are processed manually. Working conditions such as these make no allowance for any parallel processing (manual/computer aided). Comparisons made in some cases by analysis of specific subproblems (depots, specific lines) are not sufficiently informative.

- We have found by experience that the introduction of a computer-aided system is bound to bring comprehensive changes in the structure of services offered, as these turning points in planning work are used in most cases to carry out necessary restructuring of line networks and timetables. Also, relatively small changes in the services offered – arising from transitions from one timetable to the next – have a considerable effect on the number and structure of vehicle schedules, and therefore on duties.

Where qualitative aspects are concerned, the use of the HOT system brings definite improvements to working procedures and actual schedules. However, it depends from what perspective one makes the assessment (i.e. from which group: board of directors, planning staff, drivers, passengers, etc.) and for this very reason can vary from case to case. Here the following basic points apply:

- Planning time for one timetable period can be reduced by up to one-third if computer aided. This means the actuality of planning data and schedules can be increased with regard to timetable transition. At the same time it is also possible to handle additional timetable periods simultaneously. As a rule, when planned manually, there are no more than two timetable periods (summer/winter timetables). Alterations are made for seasonal fluctuations in services at a disposition level. On the other hand, companies using the HOT system work out six or more timetable periods in the course of one year (timetables for holidays, Christmas, New Year and long shopping days before Christmas).

- This means that it is possible to find a cost-effective solution for every period of time defined, or at least to establish a desirable solution which complies with company and social aspects. The latter is considered particularly important by employees, as the number of drivers able to go on holiday with their families during school holiday times can be raised owing to improved planning. As well as this, on special bank holidays either the workload can be distributed between a specific number of drivers, or duties can be arranged for as long as possible (within legal restrictions), so that as few drivers as possible are on duty.

- An additional aspect is the adjustment and/or fixing of new timetables due to short-term changes caused by extraneous circumstances – ones likely to impede operations over a fairly long period of time. These may, for example, be road works which are likely to be carried out for a long time but which have been set up in the course of one timetable period, or short-term changes in demand caused by an additional origin or destination (e.g. the opening of a new shopping centre or block of offices, etc.). Changes from manual to computer-aided processing cause drastic curtailments in work procedure, particularly in cases such as these.

- A system such as HOT is particularly suited for simulation purposes. In-company parameters can be altered for simulations (duty and break regulations, differentiated disposition of relief points, use of specific vehicle types, etc.). Analysis is then made and figures produced for how these changes affect results. For instance, running times, layovers and break regulations were varied for a specific case and this simulation was used for one level of services offered (lines, trips), so that it could be seen how they affected operations as a whole.

- Cost savings can be made in the short term, through operational cost, and in the long term, via investment requirements, if a fairly constant level of services offered is taken as a basis. Available funds can also be used for improvements in service level, so that greater efficiency is attained for the same expenditure. In many cases, however, savings take the form of compensating for constant rises in costs.

- Apart from the points already mentioned, from the drivers' point of view the system has many other positive aspects. When planning procedures are formalized, parameters have to become standardized. This generally only takes place via in-company agreements and means that planning staff have a stricter framework around which to work and planning operations are more clear cut. If this framework is consistently adhered to, duty scheduling can be structured more favourably ('fairer duties'). Computer-aided systems also produce better working documents for drivers – particularly where legibility and error avoidance are concerned – and additional instruction papers. An example of this would be an off-duty list covering a whole year, which enables drivers to make long-term plans for their private lives.

- From the passenger point of view, application of the HOT system has many advantages where presentation of passenger information is concerned. Timetables – whether timetable booklets or wall timetables – are much more up to date, better presented and include fewer mistakes. A concrete example of this is where 12 000 wall timetables were worked out within two weeks and required virtually no revising. Furthermore there are fewer discrepancies between given timetable data and the actual state of operations, because more exact planning is possible if differentiated running times are compiled according to the time of day.

Seen *in toto* application of the HOT system means that activities of the more quantitative kind are relocated to automated processing and data administration is relocated to an efficient data management system. This means considerable capacities can be released which are then available for qualitative improvements in results. Whereas with manual processing most time and expenditure was given up to establishing admissible solutions, in an efficient computer-aided system the basis of its planning activities is formed by the start solution being drawn up by an automated process.

As far as the application of sensitivity analysis is concerned (see Section 9.3.2), quantifiable evidence can be given here. In this case it has been proved that by making only slight changes in schedules and in-company buffer times (cf. Table 9.1, also Daduna *et al.*, 1993), considerable savings can be made which affect operating results as a whole. Savings on vehicle requirements are in addition to those which would arise through an isolated application of the vehicle scheduling procedure (see Section 9.3.3).

All in all, there are considerable benefits from applying the HOT system, whether for arranging planning procedures or in respect of attained results. Even if during the run-in phase some people were sceptical about it, it was possible to eliminate such feelings after further practical use of the system and a general period of familiarization.

9.6 Future prospects

Increased application of computer-aided systems will occur within public transport companies in the years to come and will be of increasing importance particularly for economic reasons. This is where the system integration mentioned in Section 9.2 will be an increasingly decisive factor. Isolated computer systems, which often have to be linked to each other via specific interfaces at a later date, and at considerable expense, do not attain the efficiency required in this case. For this reason, when choosing a system, of prime importance should be the interaction of components which are based on a relational data base in which all the necessary operational data are administered centrally.

The HOT system can meet any of these requirements over the years to come, as this chapter shows. This does not mean, however, that development

Table 9.1 Results obtained from the application of sensitivity analysis (summer 1992)

Public transport company	Number of vehicles		Savings (%)
	Start solution	Gained solution	
Hamburger Hochbahn Aktiengesellschaft (HHA)	743	735	1.1
Verkehrsbetriebe Hamburg- Holstein AG (VHH)	350	340	2.9
Dresdner Verkehrsbetriebe (DBV)	143	139	2.8
Compagnie des Transport de Besançon (CTB)	137	126	8.0

of the system has reached an end. Further developments will concentrate on introducing additional innovative hardware technologies and software components and on utilizing new and/or improved OR methods. Also interfaces have to be established to other, preceding and/or successive systems. This is of particular importance for schedule synchronization and personnel dispatching, and also for AVM systems which include, among other things, functions such as vehicle location, comparing planned timetable data and actual data, and ensuring transfer coordination.

References

DADUNA, J. R. (1988) A decision support system for vehicle scheduling in public transport, in W. Gaul and M. Schader (eds) *Data, expert knowledge and decisions*, pp. 93–102, Berlin: Springer.

—— (1992) The integration of computer-aided systems for planning and operational control in public transit, in: M. Desrochers and J.-M. Rousseau (eds) *Computer-aided transit scheduling*, pp. 347–58, Berlin: Springer.

DADUNA, J. R. and MOJSILOVIC, M. (1988) Computer-aided vehicle and duty scheduling using the HOT programme system, in J. R. Daduna and A. Wren (eds) *Computer-aided transport scheduling*, 133–46, Berlin: Springer.

DADUNA, J. R. and PAIXÃO, J. (eds) (1995) *Computer-aided transit scheduling*, Berlin: Springer.

DADUNA, J. R. and WREN, A. (eds) (1988) *Computer-aided transit scheduling*, Berlin: Springer.

DADUNA, J. R., MOJSILOVIC, M. and SCHÜTZE, P. (1993) Practical experiences using an interactive optimization procedure for vehicle scheduling, in D.-Z. Du and P. M. Pardalos (eds) *Network optimization problems: Algorithms, applications and complexity*, pp. 37–52, Singapore: World Scientific.

DESROCHERS, M. and ROUSSEAU, J.-M. (eds) (1992) *Computer-aided transit scheduling*, Berlin: Springer.

PETZOLD, P. and SCHÜTZE, P. (1993) Integrated data processing for public transport in Hamburg, Paper presented at the 6th International Workshop on Computer-aided Scheduling of Public Transport, Lisbon, July 1993.

ROUSSEAU, J. M. (ed.) (1985) *Computer scheduling of public transport 2*, Amsterdam: North-Holland.

VÖLKER, M. and SCHÜTZE, P. (1993) Recent developments of the HOT system, Paper presented at the 6th International Workshop on Computer-aided Scheduling of Public Transport, Lisbon, July 1993.

WREN, A. (ed.) (1981) *Computer scheduling of public transport*, Amsterdam: North-Holland.

CHAPTER TEN

Planning the size and organization of KLM's aircraft maintenance personnel[1]

M. C. DIJKSTRA, L. G. KROON, J. A. E. E. VAN NUNEN,
M. SALOMON and L. N. VAN WASSENHOVE

10.1 Company and problem area

KLM Royal Dutch Airlines has been the major Dutch carrier since 1919. KLM's home base is Schiphol Airport near Amsterdam. Currently (1993), KLM owns about 90 aircraft of eight different types. With this fleet, KLM carries out flights to about 150 cities in 79 countries. Of course, the safety of passengers and crew has top priority. To guarantee safety, KLM carries out high-quality aircraft maintenance. About 3000 employees of KLM's maintenance department take care of this. They also carry out maintenance operations on aircraft belonging to about 30 other carriers that have maintenance contracts with KLM.

Preventive aircraft maintenance consists partly of *major* inspections and partly of *minor* inspections. Major inspections are performed in KLM's hangars after a certain number of flight hours, depending on the aircraft type. Major inspections take several hours up to several months. The longest major inspections may involve checking all individual parts of an aircraft. Minor inspections are conducted during the ground time between arrival and departure at the airport. A minor inspection, also called a *project*, includes the following services:

- *arrival services*, which consist of fixing ground power supply, compiling a list of technical complaints based on the crew's flight records, and collecting resources (such as mobile cranes and scaffoldings) for the platform services;

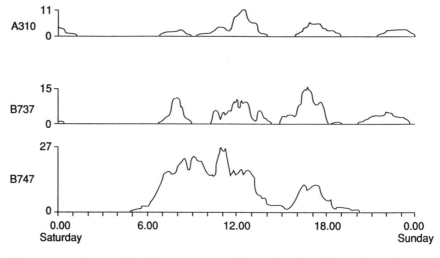

Figure 10.1 The workload between Saturday midnight and Sunday midnight for mechanical engineers contains at most 11 parallel jobs on A310s, 15 parallel jobs on B737s, and 27 parallel jobs on B747s.

- *platform services*, which consist of checking the technical state of the aircraft, and, if necessary, performing repairs; and
- *departure services*, which consist of performing a final technical check of the aircraft.

Maintenance engineers of KLM's aircraft maintenance group VOC (Vliegtuig Onderhoud Centrum, in Dutch) carry out these tasks. The workload of KLM-VOC's technical staff is mainly based on

- KLM's timetables and those of other carriers,
- contracts with other carriers, and
- maintenance standards.

The maintenance standards specify (1) in which time interval KLM must schedule each service, (2) how much time must be spent on each service, and (3) the skills required for each service (there are mechanical, electrical, and radio-skilled engineers). Usually, maintenance standards are specified by aircraft manufacturers, governments, and the carriers themselves.

KLM's timetable and those of other carriers usually have cyclical patterns, with a cycle length of one week. As a result, the workload of KLM-VOC also shows a cyclical pattern. Furthermore, the workload on an average day shows some clearly distinguishable peaks, caused by KLM's desire to limit the waiting times for transit passengers. For example, early in the morning a stream of intercontinental flights arrives at Schiphol Airport. Shortly thereafter a stream of continental flights departs from Schiphol Airport to several destinations in Europe.

Table 10.1 Each maintenance team has 20 engineers, all having a specific licence–skill combination. For example, two engineers in the team have obtained a mechanical skill with licences for A310 and B747 aircraft

	Skills		
Licences	Mechanical	Electrical	Radio
A310/B747	2		1
B737/B747	1		1
B737/DC10	1	1	1
A310/B737	1	1	
A310/DC9	1	1	1
A310	1	1	
B747	3	1	1

KLM wants to increase the utilization rate of its fleet and to smooth the workload of its ground service departments. It can accomplish the latter by increasing the number of peaks and reducing the size of the peaks. KLM-VOC's management is interested both in aggregated and in more detailed information on the workload pattern, since this provides valuable insights into the size and organization of the workforce it needs. Figure 10.1 shows a typical workload pattern.

KLM-VOC's workforce consists of about 250 ground engineers and 150 non-technical employees. The ground engineers are highly skilled and well-trained employees, since their job is a very responsible one. A governmental rule specifies that an engineer is allowed to carry out inspections on a specific aircraft type only if that engineer is licensed for that aircraft type. Besides a licence for an aircraft type, an engineer also has a specific skill for mechanical, electrical, or radio operations. The engineers obtain their licences for aircraft types and skills by attending training programmes consisting of theoretical and practical courses, field training and exams. Depending on the engineer's experience, it takes several months to several years to complete a training programme. In addition to inspections that require licensed and skilled employees, KLM-VOC carries out a small number of jobs for which no licence is required.

It would be preferable if all engineers had licences for all aircraft types and all skills. Then, they would be totally flexible. However, KLM's internal safety rules limit engineers to licences for at most two aircraft types and one skill.

The engineers operate in teams, which are KLM-VOC's smallest organizational subunits. Currently there are 12 teams of about.20 engineers each. KLM assigns engineers to teams so that the teams are almost identical with respect to their available licences and skills (Table 10.1).

The teams operate in a four-shift system with an early day shift from 06:00

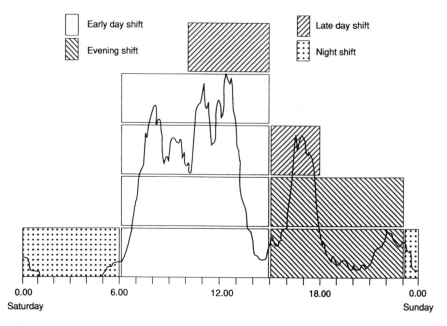

Figure 10.2 In this scheme for a typical Saturday, each team of engineers is represented by a block. For example, four teams of engineers are available in the early day shift (06:00–15:00).

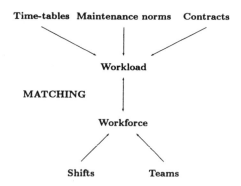

Figure 10.3 Various components contribute to the problem of matching the workforce to the workload.

to 15:00, a late day shift from 10:00 to 18:00, an evening shift from 15:00 to 23:00, and a night shift from 23:00 to 06:00 (Figure 10.2). The assignment of teams to shifts is constrained by several governmental, union and internal KLM rules. For instance, the average number of shifts per week for each team should be five, and each team should have at least one day off between a night shift and the next day shift.

10.2 Managerial problems

The managers' main problem is to find a good match between workload and workforce. The elements that play a role in this match are timetables, maintenance norms, and contracts for the workload, shifts and teams of the workforce (Figure 10.3).

The quality of the match in a certain planning period (day, week or month) is expressed in terms of the *service level* and the *utilization rate*. These performance indicators are defined as follows:

$$\text{service level} = \frac{\text{no. of maintenance jobs carried out in time}}{\text{total no. of maintenance jobs}}$$

$$\text{utilization rate} = \frac{\text{total no. of productive man hours}}{\text{total no. of available man hours}}$$

The main consequences of a bad match are the following:

- A low service level, caused by too high a utilization of the workforce. A low service level corresponds to delays, which should be avoided because they cause customer dissatisfaction and high costs.

- A low utilization rate of the workforce, caused by too many engineers, or engineers with inappropriate licences or skills being assigned to a shift. Of course, both lead to an inefficient and hence costly maintenance organization.

Planning for a good match of workload and workforce is therefore important, and it involves both the strategic and tactical planning levels. At the strategic level, KLM-VOC's management must:

- Set an appropriate target for the service level.

- Obtain insight into the relationship between the size and organization of the workforce and the resulting service level.

- Determine the impact of timetable adjustments on the size and organization of the workforce.

- Analyse the consequences of the introduction of new aircraft types on the size and organization of the workforce.

- Evaluate the consequences of partnerships with other companies with respect to maintenance.

KLM-VOC's management is also faced with a number of problems at the tactical decision level. Examples are

- To evaluate the financial implications of potential new contracts with other carriers.

- To determine the number of shifts per day and the beginning and ending times of each shift.

- To compose the teams appropriately with respect to licence combinations and skills.

- To develop a training programme, i.e. to determine the capacity and content of the educational programme needed for maintenance engineers.

- To estimate next year's personnel budgets for the maintenance department.

In dealing with these problems, the management of the maintenance department focuses in particular on the problems of coping with different aircraft types, licence combinations, and skills. The managers based their strategic and tactical capacity planning on rough aggregate workload calculations for the various aircraft types. They could not evaluate the problems caused by the different aircraft types, licence combinations, and skills to a sufficient level of detail. Furthermore, carrying out these calculations was very time consuming. KLM-VOC's management had the impression that the quality of decision making could be improved by the introduction of a decision support system (DSS).

10.3 Decision support system

The DSS consists of a data base module, an analysis module, and a graphical user interface.

The data base module stores generic data, data about the workload, and data about the workforce. The generic data include information on carriers for which KLM-VOC carries out maintenance operations and information on aircraft types. Data on the workload consist of arrival and departure times of aircraft at Schiphol, and maintenance standards. Arrival and departure times are taken directly from the timetables of the airline companies. The maintenance standards specify (1) the total time required for each maintenance operation, (2) the allowable time interval during which a maintenance operation must be carried out relative to a plane's arrival and departure time (e.g. a specific maintenance operation should not start earlier than 10 minutes after arrival and should not be finished later than 20 minutes before departure), and (3) that maintenance operations should start as early as possible. Finally, data on the workforce consist of the licence–skill combinations within a team, and the shift schedules that specify the beginning and ending times of the shifts, as well as the number of teams per shift.

A complete set of tables with generic data, data on the workload, and data on the workforce is called a *scenario*. The data base module provides functions for operations on complete scenarios and functions for operations on individual tables of a selected scenario. By analysing different scenarios the user of the DSS obtains valuable insight into the effects of changes in the data.

The analysis module provides extensive possibilities for analysing scenarios. It consists of routines for (1) estimating the workload, (2) optimizing the size and organization of the workforce, and (3) evaluating the quality of the match between workload and workforce.

The routine for estimating the workload is based on the maintenance operations and the maintenance standards. For example, consider mechanical work related to a platform inspection of a Boeing 747 on incoming flight KL-342 with arrival time 09:00 and departure time 11:25. Total ground time at Schiphol is 145 minutes. The maintenance standard for the mechanical part of the platform inspection of a Boeing 747 specifies that (1) the total time required to carry out the work is 160 minutes, (2) the operations should take place in the interval from 15 minutes after arrival to 30 minutes before departure, and (3) the operations should start as early as possible. Since only 100 minutes are available for the platform inspection, the operations are carried out by two engineers with 80 minutes of work for each. Both engineers will be scheduled to start their work 15 minutes after arrival (09:15) and will be finished 80 minutes later (10:35).

All maintenance operations carried out by KLM-VOC are translated into estimated workloads by calculations like these. Figure 10.1 shows an example of a resulting workload pattern. Owing to circumstances such as delays or failures, the actual workload may differ from the estimated workload. The influences of such irregularities on operations can be analysed by simulating these circumstances within a scenario (e.g. changing maintenance standards, or generating delays on arrival times).

The routine of optimizing the size and organization of the workforce uses the following input:

- workload estimates per aircraft type (output of the workload estimation routine) in terms of jobs with fixed starting times, fixed finishing times, and required skill;
- beginning and ending times of the shifts;
- the number of teams per shift; and
- the required service level.

Based on this input, the optimization routine calculates the number of engineers per team with a particular licence–skill combination and assigns engineers to the maintenance jobs so that (1) the total number of engineers is minimal, (2) the service level constraint is satisfied, and (3) each job is carried out by an engineer with an appropriate licence–skill combination. For example, if the service level is set to 1.00, the system computes the number of engineers per team with a particular licence–skill combination, such that all maintenance jobs are carried out and the total number of engineers is minimal. In the case of a user-specified service level of 0.98, the system calculates a composition of the teams such that at least 98 per cent of the maintenance jobs are carried out. We describe a simplified version of the

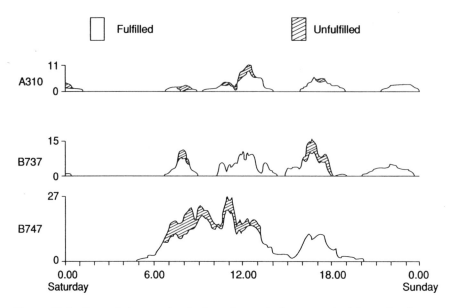

Figure 10.4 Part of the mechanical work needed on a typical Saturday for aircraft types A310, B737 and B747 can be carried out ('fulfilled') by the available engineers (white area), and part of the work ('unfulfilled') cannot be done in time (shaded area).

integer programming model used by the optimization routine in the appendix (model 1). A heuristic procedure within the optimization routine quickly generates good solutions to this model.

The routine for evaluating the maintenance processes within KLM-VOC determines the quality of the match between workload and workforce. It measures the quality of the match in terms of service level and utilization rate. The evaluation routine uses the following input:

- workload estimates per aircraft type (output of the workload estimation routine) in terms of jobs with fixed starting times, fixed finishing times, and required skill;
- beginning and ending times of the shifts;
- the number of teams per shift; and
- the number of engineers per team and their licence–skill combinations.

Based on the input, the evaluation routine assigns engineers to maintenance jobs so that the service level is maximized. The results of the procedure can be presented graphically (Figures 10.4 and 10.5). We describe the model underlying the evaluation routine in the appendix (model 2). Within the system, it is solved by a fast and effective heuristic procedure.

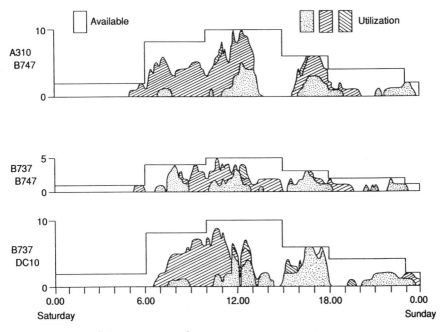

Figure 10.5   The assignment of mechanically skilled engineers to maintenance jobs on a typical Saturday for licence combinations A310/B747, B737/B747 and B737/DC10. The white area indicates how many engineers are available (at midnight one for B737/B747 and two for A310/B747 and B737/DC10, increasing to five for B737/B747, 10 for A310/B747 and 10 for B737/DC10 around midday). The other areas reflect how many of these engineers are assigned to each aircraft type: the dotted area for the engineers' first licence, and the striped area with inclining pattern for the engineers' second licence. The striped area with declining pattern indicates work for which no licence is required. For example, for engineers with licence combination A310/B747, the dotted area shows how many engineers work on A310s (first licence), and the inclining stripes show how many engineers work on B747s (second licence).

The optimization routine and the evaluation routine differ from each other in the following sense: the optimization routine considers the service level as given and determines the best composition of the teams with respect to the number of engineers required and their licence–skill combinations, whereas the evaluation routine considers the composition of the teams as given and determines the maximum number of maintenance jobs that can be carried out in time (i.e. it maximizes the service level).

Since it is not feasible to change the composition of the teams in the short run (the educational programme to obtain new licences takes several months to several years) and since in the long run management wants to design the composition of the teams so that in principle all maintenance work can be carried out (service level 1.00), KLM uses the optimization routine mainly in strategic studies and the evaluation routine in tactical studies.

10.4 System development and usage

In 1988, KLM set up a project to develop the DSS. The project team consisted of five persons from academia and five persons from KLM-VOC. At the start of the project, management had only a rough idea of the problems the DSS should handle and of the look and feel of the system. To structure the decision problems and to determine the functional requirements of the system, we adopted a prototyping approach. One functional requirement management set was that the DSS should be standalone and interface with the database systems on KLM's mainframe. Another functional requirement was that the system should run on personal computers on the employees' desks.

We developed the database module of the system using a tool for generating databases based on Clipper, and we developed the analysis module and the graphical user interface using Borland's Turbo Pascal and some tools for graphics. We devoted a great deal of effort to making the system user-friendly, which resulted in quick user acceptance. We installed the system in 1990, and KLM-VOC has used it since then.

The DSS is used mainly by staff employees to answer questions posed by KLM-VOC's managers. The system has provided valuable support in the following studies:

- a study on the impact of new timetable structures (e.g. timetables with an increased number of arrival and departure peaks per day);
- a study on the effects of allowing engineers to hold licences for three aircraft types instead of two;
- a study to estimate the workforce required per group of gates, in case maintenance processes become decentralized per gate group at some point in the future;
- a study to determine the number of engineers and their licence–skill combinations for 1992 and 1993; and
- a study on the effects of contracts with other carriers.

The DSS provides KLM-VOC's management with information that was either not available before, or was too time consuming to collect within a short planning cycle. It increases their insight into the various problems that must be solved within the maintenance department. The managers consider it a valuable tool for analysing strategic and tactical problems. The users of the system even advocated its use to other departments of KLM, such as the helicopter department.

However, evaluating a DSS is generally more difficult than evaluating more traditional information systems, since simple criteria like costs and benefits are hardly useful (Keen, 1981). First, a DSS is never completely finished and therefore costs are difficult to specify. Second, the benefits of a DSS are often largely qualitative, e.g. the impact on the organization, the quality of decision-making processes, and the resulting decisions. Evaluating these benefits in quantitative terms is difficult and has not yet received much attention in the literature (Elam *et al.*, 1986).

When considering the results of a DSS, it is important to keep in mind that they are based on mathematical models that are abstractions of reality. Optimality in mathematical terms need not necessarily match optimality in practical terms. Furthermore, most of the calculations within a DSS are based on approximation algorithms. Therefore, the results of a DSS must be handled with care. The user of the DSS must judge the practical value of a solution in the light of qualitative or quantitative considerations that were not explicitly taken into account by the models within the DSS. Hence, a DSS must be used in an interactive way, where the intelligence of the user is combined with the capability of the DSS to organize and process enormous amounts of data and to solve complex mathematical decision problems using sophisticated operational research techniques.

Acknowledgements

We are grateful to René Kalmann, Dolf Beltz, Paul Chün, Thom Grobben and Jan Smit of KLM's maintenance department for their cooperation and helpful comments during all development phases of the DSS.[2]

Notes

[1]Reprinted by permission, 'Planning the Size and Organization of KLM's Aircraft Maintenance Personnel', *INTERFACES*, Volume 24, Number 6, November–December 1994. Copyright 1994, the Operations Research Society of America and the Institute of Management Science, 290 Westminster Street, RI 02903.
[2]R. R. Kalmann, Director A310/DC-10/MD-11 Maintenance, KLM Royal Dutch Airlines, PO Box 7700, 1117 ZL Schiphol Airport, The Netherlands, writes: 'I confirm that the system described in the paper "Planning the size and organization of KLM's aircraft maintenance personnel" has proved to be a useful management tool for analysing several strategic and tactical problems that appear in personnel planning for aircraft maintenance.'

Appendix: mathematical description

We present a mathematical description of the problems solved within the analysis module. Recall the following assumptions: (A1) the number of teams is fixed; (A2) all teams have the same number of engineers with a particular licence–skill combination; and (A3) each engineer has exactly one skill.

To keep the presentation clear, we first describe the situation in which the service level equals 1.00 and in which there is only one shift and one skill. However, as Kroon (1990) and the model extensions below show, it is easy to modify the model in such a way that (1) lower service levels, (2) multiple shifts, and (3) multiple skills are properly taken into account.

We suppose that the set J of jobs (services) has to be carried out. Job $j \in J$ requires continuous processing in the interval (s_j, f_j) and is related to an aircraft of type a_j. Each engineer is assumed to have a licence combination that specifies the aircraft types the engineer is allowed to work on. The set of different licence combinations is denoted by C. Furthermore, the set J_c denotes the set of jobs that can be carried out by engineers with licence combination c. Conversely, we use the notation C_j for the set of licence combinations that can be used for carrying out job $j \in J$. Furthermore, $\{t_p | p \in P\}$ is the set of starting times of the jobs; that is, $\{t_p | p \in P\} = \{s_j | j \in J\}$.

Model 1 (optimization routine)

Here the problem is to determine the minimum number of engineers with appropriate licence combinations, such that all jobs can be carried out. Given the assumptions A1 and A2, minimizing the total number of engineers is equivalent to minimizing the number of engineers per team. The decision variables of the related integer programme are defined as follows:

X_{jc} = a binary variable indicating whether job j has to be carried out by an engineer with licence combination $c \in C_j$
Y_c = an integer variable indicating the number of engineers with licence combination c in each team.

In terms of these decision variables, the objective and the constraints of model 1 are

$$\min Q = \sum_{c \in C} Y_c \tag{A10.1}$$

subject to

$$\sum_{c \in C_j} X_{jc} = 1 \quad \text{for } j \in J \tag{A10.2}$$

$$\sum_{\{j \in J_c | s_j \leq t_p < f_j\}} X_{jc} \leq Y_c \quad \text{for } c \in C, p \in P \tag{A10.3}$$

all variables are integer. $\tag{A10.4}$

The objective function (A10.1) expresses that we are interested in minimizing the number of engineers per team. The constraints (A10.2) guarantee that each job is carried out exactly once. The constraints (A10.3) specify that the maximum job overlap of the jobs that are assigned to the engineers with licence combination c should not exceed the number of available engineers with licence combination c. This implies that a feasible solution to the integer programme can be transformed into a feasible assignment of jobs to engineers and vice versa. Finally, the integrality constraints (A10.4) specify the integer character of the decision variables.

Kolen and Kroon (1992) show that in general this optimization problem belongs to the class of NP-hard problems. Kroon (1990) and Kroon *et al.* (1993) present algorithms that can be used to find optimal or approximate solutions.

Model 2 (evaluation routine)

Here the problem is to determine the maximum number of jobs that can be carried out (i.e. maximum service level) given the size and the composition of the workforce. The prespecified number of engineers with licence combination c is known and denoted by M_c. In terms of the decision variables X_{jc} (see model 1), the objective and the constraints of model 2 are stated as follows:

$$\max Q = \sum_{j \in J} \sum_{c \in C_j} X_{jc} \tag{A10.5}$$

subject to

$$\sum_{c \in C_j} X_{jc} \leq 1 \quad \text{for } j \in J \tag{A10.6}$$

$$\sum_{\{j \in J_c | s_j \leq t_p < f_j\}} X_{jc} \leq M_c \quad \text{for } c \in C, p \in P \tag{A10.7}$$

all variables are integer. (A10.8)

The objective function (A10.5) expresses that we are interested in maximizing the number of jobs that are carried out. The constraints (A10.6) guarantee that each job is carried out at most once. The interpretation of the constraints (A10.7) and (A10.8) is similar to that of the constraints (A10.3) and (A10.4) in model 1.

Kolen and Kroon (1991) show that in general this optimization problem belongs to the class of NP-hard problems. Algorithms that can be used to find optimal or approximate solutions are described by Kroon *et al.* (1995).

Model extensions

We now briefly discuss three extensions of model 1 corresponding to (1) service levels less than 1.00, (2) multiple shifts, and (3) multiple skills.

1   *Service level less than 1.00.* In this case constraint (A10.2) is replaced by constraints (A10.2′) and (A10.2″):

$$\sum_{c\in C_j} X_{jc} \leq 1 \quad \text{for } j \in J \tag{A10.2′}$$

$$\sum_{j\in J}\sum_{c\in C_j} X_{jc} \geq \alpha |J| \tag{A10.2″}$$

where α $(0 \leq \alpha \leq 1)$ is the required service level.

2   *Multiple shifts.* In order to take into account multiple shifts, constraint (A10.3) is replaced by constraint (A10.3′):

$$\sum_{\{j\in J_c | s_j \leq t_p < f_j\}} X_{jc} \leq N_p Y_c \quad \text{for } c \in C, p \in P \tag{A10.3′}$$

where N_p is the number of teams present in the time interval (t_p, t_{p+1}). The numbers N_p are input by the user and follow from the beginning and ending times of the shifts, and from the number of teams per shift. Note that (A10.3′) is based on assumption A2, which states that teams have identical compositions with respect to licences.

3   *Multiple skills.* Since each engineer has one skill only (assumption A3), the multiple-skill problem is solved by decomposing it into several one-skill problems as formulated by model 1.

Note that model 2 can be extended to include multiple shifts and multiple skills analogous to model 1.

References

ELAM, J. J., HUBER, G. P. and HURT, M. E. (1986), An examination of DSS literature (1975-1985), In: *Decision Support Systems: A Decade in Perspective*, E. R. McLean and H. G. Sol (eds) Amsterdam: North-Holland.

KEEN, P. G. W. (1981) Value analysis: justifying decision support systems, *MIS Quarterly*, **5** (March), 1–15.

KOLEN, A. W. J. and KROON, L. G. (1991) On the computational complexity of (maximum) class scheduling, *European Journal of Operational Research*, **54**, 23–38.

        (1992) Licence class design: complexity and algorithms, *European Journal of Operational Research*, **63**, 432–44.

KROON, L. G. (1990) Job scheduling and capacity planning in aircraft maintenance, PhD Thesis, Erasmus University Rotterdam.

KROON, L. G., SALOMON, M. and VAN WASSENHOVE, L. N. (1995) Exact and
    approximation algorithms for the operational fixed interval scheduling problem,
    European Journal of Operational Research, **82**, 190–205.
    (1996) Exact and approximation algorithms for the tactical fixed interval scheduling
    problem, *Operations Research* (in press).

PART FOUR

Facilitator OR

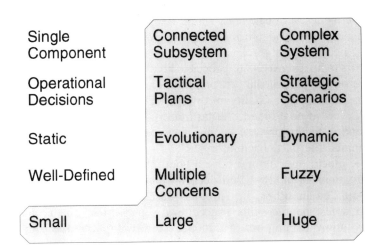

Single Component	Connected Subsystem	Complex System
Operational Decisions	Tactical Plans	Strategic Scenarios
Static	Evolutionary	Dynamic
Well-Defined	Multiple Concerns	Fuzzy
Small	Large	Huge

Fingerprints

- A complex system consisting of various highly connected subsystems. A classical example would be a supply chain where efficiency improvements at one stage may have a negative effect at downstream stages and where lack of coordination between antagonistic players may easily lead to mediocre customer service.

- Decision models are of the strategic type – though many can also be used for tactical planning. For example, supply chain models can allow for the analysis of scenarios showing the potential benefits of closer cooperation.

- The environment is often dynamic because it is during times of rapid change under competitive pressures that good scenario analysis pays off.

- Objectives are quite difficult to express because they typically deal with longer-term strategic issues, e.g. how does superior quality and flexible response affect customer retention? Moreover, there may be severe disagreement over goals and feasible scenarios.

- The data set depends on the setting. It could be rather small in some cases (highly aggregated models) but in other instances the model will need to capture all the fine detail of the situation (e.g. individual customer data).

This part shows OR's considerable potential to add value. OR can model complex interdependent parts of a system or network. In many of our industrial systems these parts have already been optimized. However, large inefficiencies often remain in the intersections between the parts due to lack of coordination. OR's power to create – or simulate – 'future reality' should put it at the highest strategic level where the money and the decision power reside. Indeed, OR models can become a powerful common language for discussion, i.e. OR as a glue.

The part requires excellent process skills as well as a broad domain knowledge on the part of the OR practitioner. Tools become less important in the sense that the personality and style of the practitioner dominate. The practitioner has to be a blend of a facilitator and a creative problem solver. Close attention to the process is one of his or her major concerns. This is why we named the part 'Facilitator OR'.

The OR practitioner should have a good feel for the capabilities as well as the limitations of model-based decision making. The models need to be solidly grounded in other, i.e. non-fact-based, business realities such as leadership styles, people skills and cultural sensitivities. The notion of an 'optimal solution' coming out of a model, so dear to many OR students, is a joke here!

This part contains four chapters dealing with planning problems in joint-product industries (e.g. dairy products), simulation of a liquid crystal display producing plant, development of an asset–liability management system for a pension fund, and a supply chain model for the potted plant sector, respectively.

CHAPTER ELEVEN

Strategic and tactical planning in joint-product industries

A Practical Application of Linear Programming

L. SCHEPENS and A. VAN LOOVEREN

11.1 Introduction

This chapter discusses a class of planning problems that are particularly well suited to be tackled with linear programming techniques: the so-called 'joint-product problems'. Joint-product problems occur in industries that are characterized by the simultaneous production of a variety of products by a process or series of processes starting from a limited number of common raw materials. No product can be produced without the accompanying appearance of the other products in the joint group.

A well-known joint-product example is the petroleum industry: starting from crude oils a variety of (joint) products like gasoline, lubricants, kerosene, naphtha and fuel oil are produced. Other examples are the meat industry and the milk industry. A meat company cannot kill a pork chop. Instead, it has to slaughter a pig, which supplies various cuts of dressed meat and trimmings. The skimming of raw milk provides not only skimmed milk, but unavoidably also an amount of cream.

In the petrochemical industry, linear programming techniques have been used for strategic and tactical planning purposes for more than 20 years. This chapter discusses the joint-product planning problems in the meat and milk industries, two environments that are less familiar with linear programming or operational research techniques.

11.2 Case 1: the meat company Belgica

11.2.1 The activities

Belgica is commercializing a wide range of meat products such as different types of smoked or boiled hams, salamis and patés. These products all use parts of the same raw materials: hams and shoulders. Some of these products use the more 'noble' or major parts, while others use the less valuable parts or byproducts, like trimmings and fat. In the first stage, the different types of hams and shoulders pass through the cutting department. The semi-finished or intermediate products from this department are subsequently treated in three production units producing several types of boiled and smoked hams, salamis and patés. Besides hams and shoulders, a number of additional raw materials are bought separately, e.g. spices, packaging materials, grease and fat.

Belgica is also able to sell the intermediate products from the cutting department, or, vice versa, to purchase them directly on the market. Given the unattractive financial conditions and the limited amounts that can be sold or purchased, this option is only used occasionally to level (temporary) production imbalances.

Belgica grew rapidly in the 1980s. At the beginning of the 1990s, the company produced more than 10 000 tons of meat products per year. Total yearly sales exceed 50 million ECU. A 'total quality' approach has led to the further development of its laboratories. Recently, a 'light' product range has been introduced in order to increase market share further. Although modern production techniques are used, the company has built up an image of traditional and artisanal production and products.

In spite of increasing sales volumes, the total profit has decreased over the last couple of years. Moreover, production imbalances have occurred frequently, forcing the company to buy or sell semi-finished products at unattractive conditions. Faced with this evolution, management decided to perform a strategic product mix analysis.

11.2.2 Cost allocation in a joint-product environment

The intention was to use the results of the analysis as a basis for making the necessary adjustments for a more profitable and balanced product mix. However, management quickly discovered that the available cost information from the cost accounting system was not very useful in providing a good insight to product profitability. This will be illustrated with a simplified case, in which only a limited number of raw materials and end products are considered.

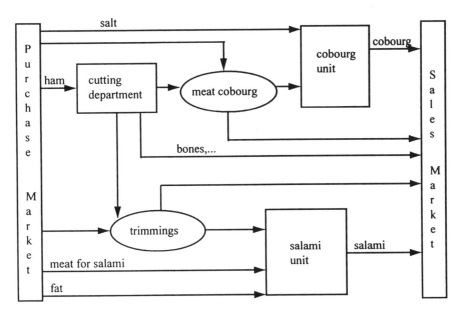

Figure 11.1   Belgica case – product flows.

11.2.2.1 A simplified case

Three production departments are considered:

1 *The cutting department.* In this department the raw material (hams) is cut into a noble product (the so-called 'cobourg'), trimmings to be used in the production of salami, and waste (e.g. bones). In reality there are several types of hams and shoulders, as well as a larger number of semi-finished products. One kilogram of ham is cut into:

  ▪ 0.530 kg of ham for the production of cobourgs
  ▪ 0.350 kg of trimmings for the production of salamis
  ▪ 0.120 kg of waste.

2 *The cobourg production unit.* Only one type of finished ham is considered: the smoked ham or 'cobourg'. To produce 1 kg of smoked ham, one needs:

  ▪ 0.680 kg of ham
  ▪ 0.320 kg of salt (brine).

  There is a loss of weight of 25 per cent on all incoming material.

3 *The salami production unit.* Again, only one type of salami is considered. Per kg, one uses:

  ▪ 0.340 kg of trimmings

Table 11.1 Belgica case – purchase prices

Raw material	Cost in ECU per kg
Ham	2.4
Salt	0.25
Meat for salami production	1.3
Fat	0.4

Intermediate product	Cost in ECU per kg
Trimmings	1.6
Meat for cobourg production	4

Table 11.2 Belgica case – selling prices

End product	Price in ECU per kg
Cobourg	6.5
Salami	5

Intermediate product	Price in ECU per kg
Waste	0.05
Trimmings	1.3
Meat for cobourg production	3.7

- 0.460 kg of purchased meat
- 0.200 kg of purchased fat.

There is a loss of weight of 30 per cent on all incoming material (maturation process).

The complete production environment is summarized in Figure 11.1. Each production unit is represented by a rectangle, each product flow by an arrow. Product flows are collected in 'product reservoirs' represented by ellipses.

The purchase prices for the different raw materials and intermediate products are provided in Table 11.1. The selling prices for the different end products and intermediate products are summarized in Table 11.2. The variable production costs (raw materials not included) are:

- cutting cost per ingoing kg of ham (mainly labour costs): 0.15 ECU per kg;
- production cost per outcoming kg of cobourg (labour, energy, packaging material): 0.80 ECU per kg;

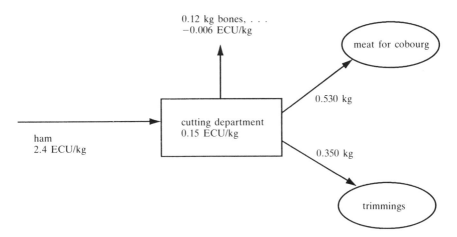

0.12 kg bones, . . .
−0.006 ECU/kg

meat for cobourg

0.530 kg

cutting department
0.15 ECU/kg

ham
2.4 ECU/kg

0.350 kg

trimmings

Figure 11.2 Belgica case – cost allocation.

- production cost per outcoming kg of salami (labour, energy, packaging material): 1.20 ECU per kg.

Based upon this information, the cost per kg produced in the cutting department can be calculated:

- purchasing cost per kg ham   2.4 ECU
- cutting cost (per kg ham)   0.15 ECU
- sales of waste (0.120 kg at 0.05 ECU per kg)   −0.006 ECU.

One kilogram of ham results in 0.880 kg of 'useful' meat at a cost of 2.544 ECU. From this 0.880 kg, 0.530 kg will be used for cobourg production and 0.350 kg (trimmings) for salami production. In order to determine the cost per kg of the two end products, the total cutting cost of 2.544 ECU has to be allocated to the two intermediate products, i.e. meat for cobourg and trimmings for salami production (Figure 11.2).

Two alternatives are considered:

- *An allocation on a physical quantity basis (alternative a).* The production cost per kg is considered to be the same for cobourg meat and trimmings. This results in a cost of 2.544/0.880 = 2.891 ECU per kg.

- *An allocation based on the 'value' of the two products (alternative b).* Assuming that the value of the trimmings equals 1.6 ECU per kg (i.e. purchase price), the resulting cost of meat for cobourg production is [2.544 − (0.350 × 1.6)]/0.530 = 3.743 ECU per kg.

At this point, all the information is available to calculate the (variable) costs for both end products (see Tables 11.3 and 11.4).

In alternative (a), cobourg costs 3.529 and salami 3.573 ECU per kg. In alternative (b) cobourg costs 4.302 (22 per cent more) and salami 2.946 (18

Table 11.3 Belgica case – cost for 1 kg of cobourg

Item	Amount per kg cobourg	Unit cost (a)	Unit cost (b)	Cost per kg (a)	Cost per kg (b)
Salt	0.32 kg/0.75 = 0.427 kg	0.25	0.25	0.107	0.107
Cutting cost for cobourg meat	0.68 kg/0.75 = 0.907 kg	2.891	3.743	2.622	3.395
Cobourg production cost				0.8	0.8
Total variable cost				3.529	4.302

Table 11.4 Belgica case – cost for 1 kg of salami

Item	Amount per kg salami	Unit cost (a)	Unit cost (b)	Cost per kg (a)	Cost per kg (b)
Fat	0.20 kg/0.7 = 0.286 kg	0.4	0.4	0.114	0.114
Meat	0.46 kg/0.7 = 0.657 kg	1.3	1.3	0.854	0.854
Cutting cost for trimmings	0.34 kg/0.7 = 0.486 kg	2.891	1.6	1.405	0.778
Salami production cost				1.2	1.2
Total variable cost				3.573	2.946

per cent less) ECU per kg. Depending on the alternative, different conclusions can be drawn with respect to the profitability of the individual products – an embarrassing and confusing conclusion!

This example clearly illustrates that in joint-product environments, product costs and product contributions cannot be calculated unambiguously. Indeed, every allocation of the joint-production costs, i.e. the cutting costs, is arbitrary.

11.2.3 The linear programming approach

Management discussed the problem with a software and consultancy company that has now implemented a financial planning software package at Belgica. The consultant has experience not only in financial modelling, but also in designing and implementing operational research applications. Based upon this expertise, it is proposed to develop a linear programming model to tackle the problem. Such a model enables the determination of the product mix that maximizes the total contribution; (arbitrary) cost allocations are no longer necessary. Moreover, product imbalances are pinpointed.

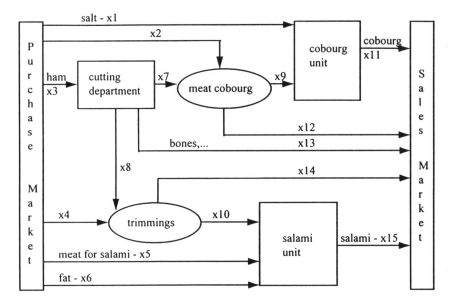

Figure 11.3 Belgica case – decision variables.

11.2.3.1 The linear programming model

In this section, the linear programming (LP) model for the simplified case of Section 11.2.2 is developed. In reality, the LP model is of medium size, containing a few thousand variables and constraints.

An LP model consists of decision variables, an objective function and constraints.

Decision variables. In the Belgica case, the decision variables correspond with the different product flows as indicated in Figure 11.3. All incoming and outgoing flows are considered. For example, trimmings are used for salami production. However, they are also:

- raw materials that can be purchased on the market;
- semi-finished or intermediate products that can be produced in the cutting department;
- end products that can be sold on the market.

This results in 15 decision variables:

- purchased products:
  X_1 amount of salt purchased on the market
  X_2 amount of meat for cobourg production purchased on the market
  X_3 amount of ham purchased on the market
  X_4 amount of trimmings purchased on the market

X_5 amount of meat for salami production purchased on the market
X_6 amount of fat purchased on the market

- semi-finished products, produced by the cutting department:
  X_7 amount of meat for cobourg production
  X_8 amount of trimmings for salami production
- semi-finished products used by the production units:
  X_9 amount of meat used in the cobourg production unit
  X_{10} amount of trimmings used in the salami production unit
- products sold on the market:
  X_{11} amount of cobourgs sold on the market
  X_{12} amount of meat for cobourg production sold on the market
  X_{13} amount of waste (bones, etc.) sold on the market
  X_{14} amount of trimmings sold on the market
  X_{15} amount of salamis sold on the market.

In the model, all production quantities are expressed in tons and all amounts in thousand ECUs.

Objective function. The objective is to maximize the total contribution, i.e. sales revenues minus purchasing costs and production costs:

$$\max z = 6.5X_{11} + 3.7X_{12} + 0.05X_{13} + 1.3X_{14} + 5X_{15}$$
$$- 0.25X_1 - 4X_2 - 2.4X_3 - 1.6X_4 - 1.3X_5 - 0.4X_6$$
$$- 0.15X_3 - 0.8X_{11} - 1.20X_{15}.$$

Costs and revenues are directly allocated to the activities that are causing these costs or are generating these revenues. No arbitrary cost allocations are required.

Constraints. Four types of constraints can be distinguished:

1 *Equality constraints*. For each product reservoir an equilibrium constraint needs to be defined, stating that the sum of the incoming flows equals the sum of the outgoing flows. There are two reservoirs, i.e. one for the intermediate cobourg product and one for trimmings.

   cobourg equilibrium: $X_2 + X_7 = X_9 + X_{12}$
   trimming equilibrium: $X_4 + X_8 = X_{10} + X_{14}.$

2 *Transformation constraints*. In the cutting department, a ham is cut or transformed into meat for cobourg production, trimmings and waste. In cobourg production, salt and meat for cobourg are transformed into a finished cobourg. In salami production, trimmings are blended with fat and other meat to produce a salami.

   cutting department: $0.53X_3 = X_7$ (meat for cobourg)
   $0.36X_3 = X_8$ (trimmings)
   $0.12X_3 = X_{13}$ (waste).

cobourg production unit: $0.75X_1 = 0.32X_{11}$ (salt)
$0.75X_9 = 0.68X_{11}$ (meat for cobourg).

salami production unit: $0.70X_5 = 0.46X_{15}$ (meat for salami)
$0.70X_6 = 0.20X_{15}$ (fat)
$0.70X_{10} = 0.34X_{15}$ (trimmings).

3 *Production capacity constraints.* The maximum capacity of the cutting department is 400 tons of incoming ham. The cobourg production unit can handle a maximum of 300 tons of incoming meat for cobourgs, the salami production unit a maximum of 1000 tons (all flows).
cutting department: $X_3 \leq 400$
cobourg department: $X_9 \leq 300$
salami department: $X_{10} + X_5 + X_6 \leq 1000$.

4 *Sales constraints.* No more than 200 tons of cobourgs can be sold:

$$X_{11} \leq 200.$$

11.2.3.2 *The results*

The optimal solution provides a total contribution of 1 877 864 ECU. The optimal values of the decision variables are given in Table 11.5. Information about the constraints is provided in Table 11.6.

The product mix includes the maximum amount of 200 tons of cobourg (sales constraint) and 700 tons of salami (production capacity constraint). More than 200 tons of trimmings have to be purchased on the market. The 'opportunity costs', i.e. marginal savings/costs when the corresponding

Table 11.5 Belgica case – optimal values for decision variables

Decision variable		Optimal value	Opportunity cost
X_1	salt purchased	85.333	—
X_2	meat for cobourg purchased	0	−0.257
X_3	ham purchased	342.138	—
X_4	trimmings purchased	220.252	—
X_5	meat for salami purchased	460	—
X_6	fat purchased	200	—
X_7	meat for cobourg production	181.333	—
X_8	trimmings for salami production	119.748	—
X_9	meat used in cobourg production	181.333	—
X_{10}	trimmings used in salami production	340	—
X_{11}	cobourgs sold	200	+2.199
X_{12}	meat for cobourgs sold	0	−0.043
X_{13}	waste sold	41.057	—
X_{14}	trimmings sold	0	−0.300
X_{15}	salamis sold	700	—

Table 11.6 Belgica case – constraint information for the optimal solution

Constraint	Used capacity	Unused capacity	Opportunity cost
Cobourg equilibrium	—		−3.743
Trimming equilibrium	—		−1.600
Cutting department (meat for cobourgs)	—		−3.743
Cutting department (trimmings)	—		−0.050
Cutting department (waste)	—		−1.600
Cobourg production (salt)	—		−0.333
Cobourg production (meat for cobourgs)	—		−4.991
Salami production (meat for salamis)	—		−4.340
Salami production (fat)	—		−3.911
Salami production (trimmings)	—		−2.626
Capacity cutting department	342.138	57.862	—
Capacity cobourg department	181.333	118.667	—
Capacity salami department	1000	0	1.438

optimal value is increased by one unit, are indicated in the last column of Tables 11.5 and 11.6 (in technical terms these columns indicate the so-called 'reduced costs' and the 'shadow prices'). These values provide interesting management information. Some examples are given below.

The opportunity cost related to the maximum sales of cobourgs (X_{11}) indicates that total contribution would increase by 2.199 ECU per kg of cobourg that would be sold above the maximum limit of 200 tons. This information can be used to evaluate a possible commercial action to boost cobourg sales. The opportunity cost related to salami sales (X_{15}) is zero. This indicates that there is no interest in organizing commercial actions to increase salami sales. In fact, increasing sales is impossible: the available capacity of 1000 tons is used entirely for the current salami production (cf. Table 11.6). The opportunity cost related to the salami capacity indicates that an extra contribution of 1.438 ECU per kg of salami could be generated. This potential benefit can be weighed against the cost of increasing the salami production capacity (e.g. through overtime or by additional investments).

In the optimal solution, Belgica does not sell trimmings (X_{14}); all trimmings produced in the cutting department are used to produce salamis. Table 11.5 shows an opportunity cost of 0.3 ECU per kg. This cost indicates the loss per kg in case one should decide to sell trimmings anyway; this would imply that 1 kg of trimmings is sold at 1.3 ECU, and that subsequently 1 kg is purchased at 1.6 ECU to restore the production imbalance! If sales prices were to increase up to 1.6 ECU per kg or more, it would be profitable to sell.

The model can run various scenarios. When no intermediate products can be purchased or sold, i.e. trimmings and meat for cobourgs can only come from the cutting department and have to be used entirely for the production

Table 11.7 Optimal values for variables – no purchases/sales of intermediate products

Decision variable		Optimal value	Opportunity cost
X_1	salt purchased	85.333	—
X_2	meat for cobourgs purchased	0	−3.050
X_3	ham purchased	342.138	—
X_4	trimmings purchased	0	+4.229
X_5	meat for salamis purchased	162.013	—
X_6	fat purchased	70.440	—
X_7	meat for cobourg production	181.333	—
X_8	trimmings for salami production	119.748	—
X_9	meat used in cobourg production	181.333	—
X_{10}	trimmings used in salami production	119.748	—
X_{11}	cobourgs sold	200	+4.732
X_{12}	meat for cobourgs sold	0	+2.750
X_{13}	waste sold	41.057	—
X_{14}	trimmings sold	0	−4.529
X_{15}	salamis sold	246.541	—

Table 11.8 Capacity utilization – no purchases/sales of intermediate products

Constraint	Used capacity	Unused capacity	Opportunity cost
Capacity cutting department	342.138	57.862	—
Capacity cobourg department	181.333	118.667	—
Capacity salami department	352.201	647.799	—

of salamis and cobourgs, the total contribution decreases from 1 877 864 ECU to 946 330 ECU (cf. Tables 11.7 and 11.8).

Salami sales fall from 700 tons to 246 tons, resulting in an unused amount of capacity in the salami production unit of 648 tons. Belgica would have to pay up to 4.229 ECU for 1 kg of trimmings! In a similar way, one could think of a scenario in which all intermediate products have to be purchased, i.e. there is no cutting department. In this scenario, the joint-product problem would no longer exist. Unfortunately, this scenario is not at all realistic. Intermediate products can only be sold or purchased at reasonable conditions in limited amounts. Therefore, as long as Belgica remains in the meat business, it will need a cutting department and management will be confronted with the (dynamic) problem of organizing production and sales towards a well-balanced product equilibrium in order to guarantee profitability.

11.2.3.3 *Fixed costs, product costs and transfer prices*

Although the proposed approach solves the problem of arbitrary cost allocations and provides a useful tool to simulate quickly the impact of various scenarios, the following embarrassing questions remain unanswered:

- What about the fixed costs?
- Can one determine the cost for individual products?
- Is it possible to calculate a transfer price for the intermediate products?

Fixed costs: In the LP model, only variable costs and revenues are taken into account. Fixed costs have not been considered up to now. The usual approach would be to allocate fixed costs to the different product flows, in more or less the same way as the joint-production costs. This would bias the optimal solution, some products will be judged to be unprofitable because their selling price will be inferior to their (arbitrary) 'total cost', other products will seem interesting while in fact they are not. Another allocation of fixed costs could lead to other 'victims'. A more proper approach is to consider the fixed costs as a separate charge that is not immediately affected by the decisions considered in the model. This approach is simple and pragmatic, but can prove to be a bit too simplistic in certain cases. Fixed costs always depend on the complexity of the organization, e.g. the number of different products, the variety of markets and distribution channels, the diversity of the production processes. The more complex the organization, the higher the fixed costs tend to be. Methods such as 'activity-based costing' (ABC) aim to clarify the links between the activities and the fixed costs.

Belgica decided, in parallel with the product mix study, to perform an ABC study of the fixed costs in order to get a better insight of these costs and how to reduce them. One of the conclusions of this study was that the number of sales offices in different countries was an important cost driver.

Product costs: As illustrated in Section 11.2.3.2, the result of the LP model indicates which products are interesting and which ones are less interesting. However, individual product costs that could directly be used for price-setting purposes, for example, are not provided. Having such costs would imply that relationships between products could be neglected and that no arbitrary cost allocations would be needed.

Transfer prices: The opportunity costs associated with the product reservoirs indicate the loss of contribution if one unit of the corresponding product is lost. The opportunity cost (cf. Table 11.6) for the cobourg reservoir is -3.743 ECU per kg; in other words, the loss of 1 kg of meat for cobourgs would result in a lower production of cobourgs and a net loss of 3.743 ECU.

The opportunity cost for the trimmings reservoir is -1.6 ECU per kg, indicating that an extra kilogram would need to be purchased (at a cost of 1.6 ECU) in case a kilogram is lost in the reservoir.

For a given solution, these costs can be considered as 'transfer prices' between production units. However, these transfer prices are only valid for a particular solution. They will change when the product mix or production volumes change (e.g. in the no purchase/sales scenario, the opportunity costs become 0.950 and -5.829 ECU per kg!). Once again, if transfer prices could be determined once and for all, the problem could be split into independent subproblems. In that case, the transfer price for trimmings could be used as the purchase cost for the salami production unit and the planning of this unit could be organized without considering the other production units. This approach is tempting but false.

Belgica is using the model for strategic planning and budgetary purposes as well as at the tactical planning level.

11.2.4   Strategic usage of the LP model

In the case of Belgica, strategic planning covers a planning horizon of up to three years. Strategic planning focuses on:

- investments and disinvestments
- the global equilibrium between the product flows
- the impact of new product families.

Based upon new developments, trends and evolutions of products, markets and production technology, several scenarios can be defined and solved with the model. The results assist management in finding answers to the following questions:

- Are the actual and/or planned production resources sufficient?
- Which production units are or will become bottlenecks?
- Which production departments are running below capacity?
- What is the impact of purchase limitations?
- Which end products hit their sales limit and what is the extra contribution when extra sales could be realized?
- Which end products are economically not interesting and what is the loss when these products are sold anyhow? What price increases are necessary to make these products attractive?
- Is the 'make/buy' strategy with respect to the intermediate products in tune with the product equilibrium?
- Is it useful to think of new products or new valorizations of semi-finished products which are actually sold on the market?

- Is it useful to sell (some) semi-finished products on the market instead of processing them further?

The answers to these questions can lead to new and challenging ideas and can be the basis for additional scenarios.

11.2.5 The LP model as a budgeting tool

Belgica also uses the LP model to develop its yearly budget. The budget cycle starts by collecting the sales forecasts. This commercial plan contains sales quantities per product (families) and per market as well as the sales prices. The traditional 'trial-and-error' approach, involving numerous spreadsheet simulations in order to reach a feasible solution satisfying production and supply constraints, has been replaced by an optimization approach.

Based upon the LP model an optimal production and supply plan is calculated in line with the commercial plan. It is clear that this plan is only a starting-point for further discussion resulting in modifications of the commercial plan, additional simulations, etc. Based upon the production plan and the corresponding resource utilization, a profit and loss statement, a balance sheet and a liquidity plan are established.

11.2.6 Tactical usage of the LP model

Finally, the LP model is also used for tactical planning purposes, i.e. the production planning for the coming 26 weeks. The production cycle for salamis and for cobourgs takes several weeks because of the long maturation times (approximately six weeks). Given these long lead times, the purchase price fluctuations, the temporary shortages on the purchase market, and the seasonal sales pattern, it is not an easy task to develop a plan to satisfy all constraints and preserve or restore the product equilibrium.

The tactical planning model provides the following information:

- the weekly production quantities for every semi-finished product and end product;
- the weekly amounts of raw materials or intermediate products to be purchased;
- the weekly amounts of intermediate products to be sold;
- the weekly amounts of semi-finished product to be stored (in freezers).

This weekly planning model is a multi-period model considering individual products instead of product families. The tactical model is therefore much larger than the strategic or budgetary models (thousands of variables and constraints instead of hundreds).

The multi-period model can be seen as a concatenation of several single-period models (Figure 11.4). The product reservoirs of these single-

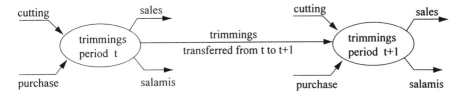

Figure 11.4 Belgica case – multi-period model.

period models are linked through 'stock flows'. These stock flows or inventory variables indicate the amounts that are stored in period t and transferred to period $t + 1$.

The multi-period model will determine the stock levels for the different products in each period starting from the initial inventories.

The usage of the three models (strategic, budgeting and tactical) is very different. The tactical model is used every week, the other models only once or twice a year. For the tactical model only one scenario is considered, but for the other models numerous scenarios are analyzed.

11.3 Case 2: the milk company Fromico

11.3.1 Activities

Fromico is a cooperative with several thousand dairy farmers as members. The company has different milk factories, each producing its own specific end products. Factories are grouped into business units. These business units are:

- *Cheese.* This business unit includes approximately 15 factories producing different types of cheese. The business unit Cheese has one central warehouse, with an immense cheese store and different packaging lines. The cheese from the cheese factories arrives in this central warehouse and is stored for a short or a long period of time (from a few days or weeks up to a few years for old cheese). The cheese is then sold as such or is first sliced and packaged on one of the packaging lines.

- *Industrial products.* This division produces and sells raw materials, semi-finished and finished goods for infant foods, dietary foods and the food industry in general. The main product families are milk powders (fat milk powder, skimmed milk powder etc.) and whey powders. All of these products are produced on three powder towers. Whey is a derivative of cheese and is bought from the Cheese division. This business unit also sells butter and butter-fat, produced in three factories. Butter and butter-fat are byproducts of cream. The inputs come from the powder installations, or are bought from the Cheese or Consumer Products divisions.

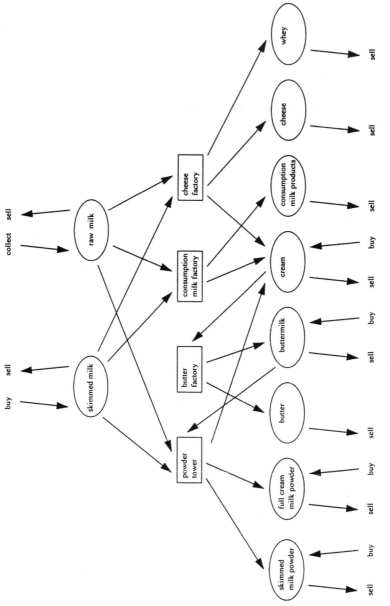

Figure 11.5   Fromico case – product flows.

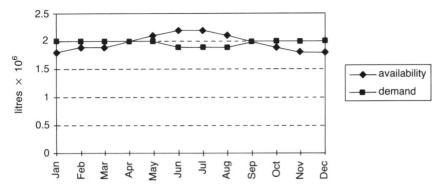

Figure 11.6   Fromico case – milk availability and demand curves.

- *Consumer products.* This unit includes one factory, producing hundreds of different consumer products. Both the Fromico brands as well as several private labels for large customers are produced. The products range from fresh milk to buttermilk and coffee milk, some of these products being available in different packagings and flavours.

A simplified and schematic overview of the product flows is given in Figure 11.5. The ellipses represent the different product groups, the rectangles the different processes.

In the past, difficulties were encountered in balancing the availability or supply of milk with the 'demand' of the different business units, each business unit wanting the milk from the farmers at the lowest possible price, and trying to convince management that *its* added value to this milk is more important than the added value of the other business units. The problem is amplified due to the seasonality of the so-called milk-curve, which is anti-cyclic to the demand curve (see Figure 11.6). Moreover, the composition (fat, protein and dry matter contents) of the milk changes throughout the year, which makes the problem even more difficult.

There are a number of ways to solve a (temporary) disequilibrium between supply and demand:

- *Extra buying.* Fromico is able to buy products on the market. These extra purchasing possibilities not only include raw milk, but also semi-finished products such as cream, skimmed milk, buttermilk, or different types of powders. Even cheese can be purchased. However, these purchases are normally committed in long-term contracts. An important challenge for management is to estimate, as correctly as possible, the timing, volume and price of these long-term contracts.

- *Selling.* In a similar way, Fromico can sell its raw material (raw milk from the farmers), or semi-finished products, to the market instead of processing them.

- *Storage.* Some of the finished products and semi-finished products can be produced in advance. Milk powder, for example, can be produced a few months before it has to be delivered. Also, in cheese production, it is possible to use milk powder as a substitute for fresh milk (up to a certain percentage). In this way, surplus milk quantities from the end of the summer can be passed over to winter production and/or sales. It is clear that this alternative causes extra production and storage costs.

11.3.2 The joint-product problem at Fromico

Fromico is confronted with a joint-product problem that is very similar to the Belgica case. This will be illustrated by the following example. Suppose that there are only two factories: a standardization factory and a powder tower. The standardization factory performs a very simple transformation: from 1 kg of raw milk it generates 0.8 kg of skimmed milk and 0.2 kg of cream. The powder tower transforms 1 kg of skimmed milk into 0.1 kg of skimmed milk powder. The company has contracts to sell 8 kg of skimmed milk powder *and* 20 kg of cream at the prices indicated in Figure 11.7 (2 ECU/kg for skimmed milk powder, 1.6 ECU/kg for cream).

The contributions of the two products, calculated via allocation of the joint-production costs on a physical quantity basis, are shown in Table 11.9. Although the total contribution is positive, i.e. 9.40 ECU, one could conclude that the production of skimmed milk powder is not an interesting activity. However, a cost allocation based upon the sales value of the products would result in a completely different conclusion. Once again, this example shows that any allocation of joint costs is arbitrary and can result in false conclusions and decisions.

For example, suppose that the company can also buy its semi-finished products on the market. Skimmed milk can be purchased at 1.8 ECU/kg and cream at 1.35 ECU/kg. Table 11.10 indicates the contributions when all intermediate products are purchased.

It is clear that this alternative is not attractive (total contribution decreases from 9.40 ECU to 5.00 ECU). However, by looking only at the contribution of skimmed milk powder, one could erroneously decide to go for a purchasing strategy. As in the case of Fromico, it was decided to use LP to get a better insight into product profitability and product (im)balances. The models are used at the strategic level (single-period model) and at the tactical level (multi-period model).

11.4 Summary

This chapter has introduced the cost allocation problem in joint-product industries and shown how it can be tackled using linear programming. Examples from the meat industry and the milk industry have been provided

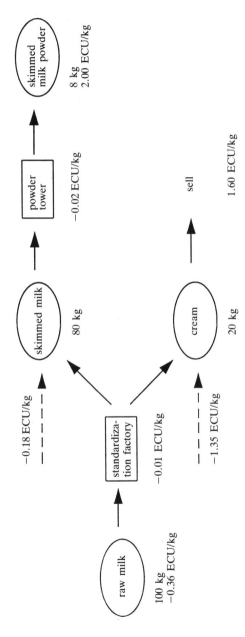

Figure 11.7 Fromico case – cost allocation.

Table 11.9 Cost allocation on physical quantity basis

Item	Skimmed milk powder (ECU)	Cream (ECU)
Sales value	$2 \times 8 = 16$	$1.6 \times 20 = 32$
Specific variable costs	$-0.02 \times 80 = -1.6$	—
Purchase cost raw milk	$0.8 \times 100 \times -0.36 = -28.80$	$0.2 \times 100 \times -0.36 = -7.20$
Variable cost standardization	$0.8 \times 100 \times -0.01 = -0.80$	$0.2 \times 100 \times -0.01 = -0.20$
Contribution per kg	-15.20	24.60

Table 11.10 Contribution when intermediate products are purchased

Item	Skimmed milk powder (ECU)	Cream (ECU)
Sales value	$2 \times 8 = 16$	$1.6 \times 20 = 32$
Specific variable costs	$-0.02 \times 80 = -1.6$	—
Purchase cost	$80 \times -0.18 = -14.40$	$20 \times -1.35 = -27.00$
Contribution per kg	0.00	5.00

to illustrate the approach. These examples are all based upon several implementations in these industries. A typical implementation effort amounts to several weeks of consultancy for problem analysis, modelling and scenario analysis. Besides the (external) consultancy effort, a lot of (internal) time is spent in gathering data. This process can be very time consuming and difficult, especially in cases where several business units or factories are involved, each unit having its own information system. Whereas the investment cost can easily be calculated (typically 2 to 3 million Belgian francs, including hardware, software and consultancy), it is much harder to measure the benefits or savings. In several cases, the proposed approach was used to prepare decisions that were of vital and crucial importance to the company, e.g. decisions related to the closure of factories or the acquisition of new companies. In general, the models are used to prepare decisions of strategic and tactical importance, having an impact on results largely exceeding the investment cost. Finally, in several cases, the – sometimes unexpected – optimization results gave rise to creative and stimulating discussions within the company resulting in, for example, the introduction of new products providing a more balanced product mix.

Bibliography

The following references provide articles and descriptions upon which the above case studies are based. These are: (1) the joint product problem in the lime industry; (2) the Belgica case (in part), and (3) the joint product problem in the refinery industry.

(1) RUTOT, D. (1984) Application de la programmation linéaire à l'industrie extractive et chafournière, Master's Thesis, Facultés Universitaires Notre-Dame de la Paix, Namur.

(2) SCHEPENS, G. and LOUVEAUX, F. (1994) Produits joints: Etude de cas Sanzot. Course notes, Facultés Universitaires Notre-Dame de la Paix, Namur.

(3) WILLIAMS, H. (1985) *Model Building in Mathematical Programming*, 2nd Edn, John Wiley: New York.

CHAPTER TWELVE

The use of simulation in the design of a production system for liquid crystal displays

H. A. FLEUREN

12.1 The environment

In this section we will give a brief view of the product, the project team and the start of the project. We discuss the production of active matrix liquid crystal displays (AM-LCDs). AM-LCDs can be produced in colour and black and white. Very many applications have been foreseen, especially for multimedia. The main advantages of AM-LCDs are the speed and the contrast ratio of the screen. These are also the main differences from the traditional (passive) LCDs which can be found in watches and electronic notebooks. Several screen diameters, from 2 inches up to more than 10 inches, (5 cm to 25 cm) can occur.

An AM-LCD consists of two substrates of glass. One of the substrates, called the active plate, contains the pixel elements. Each pixel element has its own switch (diode or transistor). The other substrate, called the passive plate, contains a colour filter and the counter electrode for each pixel. The two plates are joined by a seal at a distance of approximately 5 μm and the coupled plates are cut into cells and filled with liquid crystal. The cells are closed and finished for the final display.

The production system for AM-LCDs can be seen as consisting of three coupled factories as indicated in Figure 12.1. Processes in the active and passive parts are similar to semiconductor manufacturing, i.e. to a high precision in a clean-room environment. The cells are assembled in a process way and finishing is a typical assembly process.

Within the company of a large components manufacturer, a project team was formed with the goal of designing a complete high-volume flowline for

203

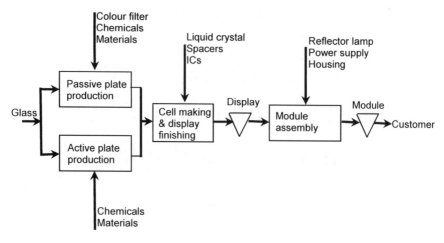

Figure 12.1 Simplified production of AM-LCDs.

the production of AM-LCDs. Several screen diameters had to be produced. This implied the design of the complete technical system, the (logistic) way of working and the complete organization. This was a rather complex task since both the product and the technology required to produce it are relatively new.

12.2 The project

The original project team was set up in 1988 with people from several business disciplines (marketing, logistics, production, etc.) to start a *feasibility* study for such a production (flow)line. Determining feasibility implied that a rather detailed design of the factory and organization had to be developed to gain an insight into the possible capacity of the factory (turnover) and the associated costs. The estimated market demand was an important but very uncertain input. Therefore feasibility at several levels of capacity had to be investigated.

After the feasibility study the go/no-go decision was taken, fortunately in favour of 'go'. The outlined feasibility study, and the detailing phase after the crucial 'go' for this complex project, took several years. During these years the project team worked on the feasibility of processes, a search for equipment manufacturers, the development of some equipment within the organization and of information systems, setting up the organization, preparation of the building, etc. In early 1993 the line started production of the first test lots.

At the start of the project, the project team was responsible for the smooth flow of products between the processes, i.e. the logistic design of the production line. Our bureau has much experience in similar projects in the

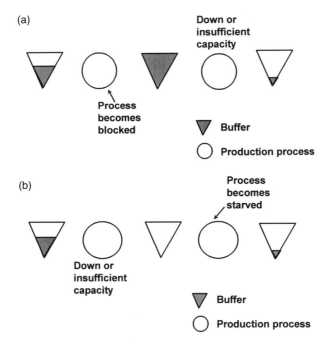

Figure 12.2   Blocking (a) and starvation (b) in a production line.

integrated circuit (IC) industry and that was why it was consulted. The bureau was familiar with OR techniques, especially with simulation, and was convinced that these techniques could help gain an insight into the logistic behaviour of the production line at an early stage of the project. From our experience in simulation projects, we know that this is one of the rare cases in which simulation is used very early in the design of a production system.

In what follows we will concentrate on the task for which our bureau was consulted: support in the logistic design of the production line to be built. To avoid confusion we will refer to our client as 'the project team' because in several phases of our support project we cooperated with several 'clients'. The final responsibility has always resided with the leader of the processes and equipment part of the project.

12.3   The problem

In this section we will briefly discuss the problem from a more technical point of view. The description is not complete, for a massive number of details and exceptions play a role in such a project. However, for the purpose of this chapter the main idea will suffice.

As outlined in the previous section we were involved at the beginning of the design of the production line. Typical questions that arose at that time were:

- What is the degree of utilization of the equipment? The answer was important for the amount of equipment installed per step.
- How large should the buffers be between processing steps? The answer was important for layout and capacity reasons. The latter will be explained below.
- What are the average contents of the buffers (work in process (WIP))? The answer was important to get a good idea of the capital tied up.
- What should the batch size be (in transport as well as in equipment)? Ideally it should be one, as we know, but what should it be in a practical environment with limitations on (chemical) processes, equipment, etc.? The answer is important for achieving a certain degree of efficiency and for balancing the production line.

We will describe the buffering problem in somewhat more detail; see also Figure 12.2. In the flowline there are many sequential production steps. Two production steps are separated by a buffer of limited size which has a number of effects:

- the buffered amount will never exceed the size of the buffer;
- when the buffer reaches its limit, the preceding process will have to stop (this is called 'blocking');
- when the buffer is empty, the following process will have to stop (this is called 'starvation').

Buffers are needed to cope with the dynamics of the line (batch sizes, disturbances, differences in cycle times, etc.). When the buffers are too small this will result in a lot of blocked and starved time for the equipment, which will decrease the capacity of the entire line. When the buffers are too large this will result in many products in the buffers (a high work in process) which results in long throughput times through the production line. Moreover, large buffer contents may lead to quality problems (deterioration) and have the risk of becoming obsolete when severe errors are detected in the previous steps. A trade-off between all these effects had to be made. This is possible with a quantitative approach.

   The dynamics of the line are influenced by a number of design parameters such as batch sizes, amount of parallel equipment, control rules, etc. Normally these parameters are more or less 'fixed' for already existing production lines. In an early design stage (as in our project), they can be used for investigating the future behaviour of the line and therefore have consequences for the design of the line.

   Our first ideas of the design of the line were based on Goldratt's bottleneck theory. The bottleneck should have enough buffering capacity in front to keep it going under many circumstances. In our case of course there should also be enough buffering capacity behind the bottleneck due to the risk of blocking.

For the entire production line, the project team tried to 'balance' the line with regard to capacity, a very common activity in designing a new line. Balancing means that the capacities of subsequent processing steps are more or less in line with each other. There was a target capacity, and equipment engineers tried to meet this capacity in their designs. They could do this by considering additional parallel equipment, adjusting cycle times within certain limits, adjusting production batch sizes, etc. All equipment engineers did their job very well at this point, which yielded interesting results in our first runs! We will come back to this below.

12.4 The OR approach

Both the project team and our bureau believed that the questions outlined above could be answered by discrete simulation techniques. There were several reasons for this choice. First of all, simulation is very suitable for incorporating the level of detail that is required to describe some aspects of the production system (think of all kinds of control rules). The alternatives for simulation are analytical methods, but our experience is that simplifying assumptions have to be made to be able to solve the analytical equations. By using them we would not win the project team's confidence and this is the second reason for choosing simulation: the team wished to recognize its suggestions and information in the model. A third reason was that some people in the project team knew about simulation from previous projects and were convinced of its value. A fourth reason for using simulation is the possibility of animation. Animation is the visualization of the production system under study with all of its dynamics. At the start of our project animation was not considered to be a serious factor. We thought of it as a toy for consultants and a demonstration tool for management. At first, therefore, we decided not to put any effort into it.

The questions to be answered were clear. The greatest difficulty in the start-up phase was to agree on the required level of detail in the study. In a model not all real-world aspects can or should be taken into account. It was sometimes difficult to convince people that their everyday problems were not important with regard to the simulation.

We agreed to start our support for a small part of the line, called the subline for the moment, and not for the entire line. This subline was important for two logistic reasons: at first it contained the predicted bottleneck, and the subline occurred several times in the flowline. Moreover, the subline was important for testing the cooperation, the way of working, etc. The intention was, if our contribution was considered successful (which fortunately it turned out to be), to answer the design questions for the entire line.

For a simulation study much data need to be gathered. In this respect, one

Table 12.1 Example of status data in project

Process	Code	# Equipment	Cycle time (s)	Throughput time (s)	Batch size	Uptime
Process aaaa	AAAA	1	30	30	1	95%
Process bbbb	BBBB	**4?**	1300	3900	10	92%
Process cccc	CCCC	2	55	110	**1?**	93%
Process dddd	DDDD	1	25	25	1	92%
Process eeee	EEEE	1			4	99%
Process ffff	FFFF	5	140	140	1	**99%?**
Process gggg	GGGG	2	**65?**	65	1	91%
etc.	. . .	. . .	. . .	. . .	. . .	. . .
. . .	. . .	. . .	. . .	. . .	. . .	. . .
. . .	. . .	. . .	. . .	. . .	. . .	. . .

can think of cycle times and throughput times of equipment, disturbance rates, production and transport batch sizes, the specification of priority and control rules, etc. One difficulty at the beginning was that the processes and their characteristics were new to us. Only after some time did we become familiar with them. In this case we had many processes each with its own equipment engineers. A problem therefore was that these equipment engineers regularly modified their processes on technical grounds. Of course this is understandable from their design tasks but many of these modifications had far-reaching consequences in terms of logistic behaviour.

A number of sessions were organized to discuss all kinds of details and the data gathering. Data collection was a joint operation between the project team and us. Owing to the many processes and the difficulty for us of knowing who was responsible for the different processes, the project team leader gave overall responsibility to one person for this task. This appeared to work much better. We presented overviews on missing and questionable data, from which it was easy to see where any gaps in the data occurred and which data were questionable. In regular meetings a table was compiled and updated with new information from the equipment engineers. An example of such a table is shown in Table 12.1. The bold labelling indicates a questionable data and no entries indicate gaps in the data.

In addition we started discussions on the scenarios that had to be worked out. Several capacity scenarios with different production systems, design and control were evaluated. One should realize that in the design stage of a production line there are many parameters that can be modified. The combination of all these parameter levels leads to a potentially enormous number of simulation runs. Good design of experiments might reduce this number drastically but the number of parameters remains large (order of hundreds). Finally a selection of starting runs was made.

12.5 Runs and validation

When building models, it is good practice after the model has been verified to make a validation run for logical and programming errors. The validation run, with parameters set as much as possible to describe the existing system, should also show a performance more or less similar to the real performance. All large differences have to be explained.

However, in this situation we did not have an existing system. Therefore the results of the starting runs were discussed extensively with the project team in order to give it confidence in the results of the model. When the results did not correspond with expectations detailed explanations were necessary. Sometimes this was difficult. It was at this time that we decided to make a simple animation, and it appeared to work very well. People obtained insight because they saw what was happening in the line, i.e. which buffers tended to become full, which buffers tended to become empty and how production batching influenced the behaviour of the line. After this phase most people agreed and accepted the results and other runs could be defined and made.

12.6 Results

As described above the final *real* capacity of the line was important together with the WIP/throughput time. In each run we used a set of design parameters and a certain buffering level. One of the most important requirements of the project team was the kind of pictures as shown in Figure 12.3. Each point in the picture represents one simulation run. By making these pictures for several product types, we obtained a good idea of the sensitivity of the line to certain parameters.

As mentioned earlier the first results were somewhat surprising: for one product type rather high capacity losses were found. At first we thought this was due to a programming error in the model but after more discussion we found the cause: the line balancing job for this particular product type had been done with the target capacity of the line. This resulted in a line with many (near-)bottlenecks. Using Goldratt's rule that one hour lost on the bottleneck is one hour lost on the production line, we could explain the results very well. Many bottlenecks consequently implied high losses for our production line.

The results seemed to imply that balancing was not a good thing to do, which was rather astonishing. On the other hand, not balancing would imply too high an investment in equipment. We could show relatively simply that, with the help of the simulation model, balancing should be done but not too tightly: a small overcapacity (10–20 per cent) of non-bottlenecks gives a much better line performance.

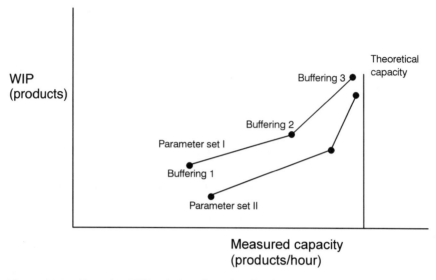

Figure 12.3 Capacity–WIP relation for LCD flowline.

12.7 State of affairs

On the basis of these pictures and several qualitative arguments the project team decided on which situation it preferred. This was a discussion-intensive process where qualitative arguments on standardization, quality management, etc., played an important role. Finally, a decision was made on which recommendations would be implemented in the line.

The production line has now been realized. We have both the physical production line and a model describing its behaviour. The shopfloor control system may yield a lot of data on cycle times, disturbance rates, etc. This information can be inserted into the model and the behaviour predicted. Such questions can be answered as:

- How will the number of product carriers in the line be influenced by other disturbance rates?
- What is the effect of operator reaction time on final capacity?

12.8 Conclusions

On evaluating the project subsequently we determined the following aspects. It is difficult to measure improvements as in an optimization problem where we can say that we saved a certain amount of money. We are convinced that certain recommendations have led to other investment decisions such as the amount of equipment installed. A more important result from an operational point of view is a more controlled production line. Short production lead times and a lower WIP are the main results, but also lower variation in lead times.

Perhaps the most important result of the simulation project is that, although we did not aim for it at the beginning of the project, many people have worked with the same model using data acceptable to almost everyone. The model and its visualization gave people a feeling for the future. We think that this has contributed much to the logistic way of thinking and working in a project team.

Appendix: the consultancy bureau

CQM is an independent consultancy firm providing services to industry and government. CQM aims to provide better control of technical as well as organizational processes in research, development, manufacturing and distribution. The CQM contribution is based on expertise in the fields of:

- statistics and quality engineering;
- optimization and simulation;
- project management in R&D;
- decision analysis and strategic modelling;
- material handling and lay-out planning.

Quantitative models and methods are ideally suited to gain insight into situations which at first glance appear complex and uncontrollable. This insight opens the way to the appropriate improvement measures.

Examples of application areas where CQM is active, are:

- process control, technically as well as logistically;
- design and development of products and processes;
- systems tolerance, systems reliability;
- design and control of distribution networks;
- optimizing usage of raw materials;
- short- and long-term production planning;
- production and stock allocation;
- vehicle planning and routing;
- lay-out problems;
- design of warehouses.

CQM was founded in 1979. The company employs 40 persons. All CQM consultants have a technical or scientific background, and they consider it as a challenge to translate their specialist know-how into practical results in close cooperation with the client.

amount of agreement of literature. A more direct treatment is an analytical approach and a more economical. Experimentally Short-period oscillations and a study will assist in testing the possible variance in both.

CHAPTER THIRTEEN

Decision support of asset–liability management

C. G. E. BOENDER

13.1 Environment and clients

The projects which are discussed in this chapter concern asset–liability management (ALM) for pension funds. In Section 13.2 we give background information about pension funds and a treatment of ALM. The total value of the investments of Dutch pension funds amounts to more than 400 billion Dutch guilders. One of the key issues of ALM is the determination of the amount of risky securities in these investments: if pension funds increase the amount of risky investments (stocks and real estate), at the expense of fixed-income securities (bonds and loans), they can suffice with lower average contributions, at the expense of higher fluctuations of the contributions and a higher probability of deficits. Dutch pension funds in the past mainly invested in fixed-income securities, such that the contributions can be reduced by increasing the investments in stocks and real estate. This currently gains relevancy due to the slowing down of the national economies and the ageing of the population. On the other hand there is a strong countercurrent due to the responsibility of pension funds to prevent the occurrence of deficits. The stake of the current and countercurrent being more than 400 billion guilders is one of the explanations of the enormous, and still growing, interest in ALM.

ORTEC started its activities in this field in 1985 for the pension fund of a bank with about 50 000 trustees and 3 billion guilders of investments. A detailed tailor-made system was developed, not for purposes of ALM, but to support the transition from a final pay pension scheme to an average pay scheme, and to support the conversion to a so-called 'dynamic funding system', introduced in the Netherlands by Brans & Co. actuarial counsellors. Later we were allowed to use the tailored software for other pension funds. To accomplish this we invested twofold. First of all the software was generalized such that it could serve as a basis for the efficient development of tailored versions for other pension funds. Second, we developed and implemented the ALM models which are described in Section 13.3, using the knowledge and experience of practitioners and researchers from the previous pension fund, an insurance company, a fund manager, and our colleagues at the Erasmus University Rotterdam and ORTEC.

Currently several versions of the decision support system are used by about 10 pension funds, which insure more than 1 million trustees, with an investment portfolio of more than 50 billion guilders. Also, versions of the system are used by an insurance company, a fund manager and the actuarial department of a management consultancy office to serve their pension fund clients. Our experience in developing and applying these systems is the content of Section 13.4. Some conclusions and recommendations are contained in Section 13.5.

13.2 The problem

13.2.1 Background

Asset–liability management for pension funds (ALM) is the problem tackled by the decision support system. To give some background information, we first consider a typical formula of a benefit-defined pension scheme:

$$
\begin{array}{l}
\text{earned rights } O_t \text{ of old-age pension} \\
\text{by trustee A in year } t
\end{array}
=
\frac{(0.7/40) \text{ (salary of A in}}{t - \text{public pension payment).}} \quad (13.1)
$$

If trustee A resigns, the yearly amount of old-age pensions payments (excluding indexations) that A will receive from the age of retirement onwards will be equal to the addition $\Sigma_{\tau \le t} O_\tau$ of the rights earned in the successive years of service. In an exemplary situation A starts building up pension rights at the age of 25 until the age of retirement of 65. Including the public pension payments which are granted by the Dutch administration to all citizens who are at least 65 years of age, A will then receive a pension of 70 per cent of average salary over the period of service. Therefore a pension scheme which is based on (13.1) is referred to as an average pay system, as opposed to a more expensive final pay system, where the pension payments are based on the level of the last received salary. Given the so-called complete funding policy the yearly contributions to be paid by the plan sponsor are equal to the expected net present value of the additional earned rights:

$$
r \text{ per cent complete funding contribution} = O_t \sum_{\tau=65-a_t}^{\infty} P_\tau/(1+r)^\tau \quad (13.2)
$$

where r is the discount interest rate, a_t is the age of A in year t, and P_τ is the probability that A is still alive τ years from today. Analogously the r per cent premium reserve is defined as the expected net present value of the total earned rights $\Sigma_{\tau \le t} O_\tau$. If the value of the investment portfolio is less than the total r per cent premium reserve of the complete insured population a deficit, or underfunding, is said to occur. A positive value between the value of the investments and the premium reserve is referred to as a surplus.

Special attention has to be paid to the consequences of inflation. Let i_t denote inflation in year t. The total of the earned rights of the trustees is usually indexed to $(1 + i_t)\Sigma_{\tau \le t} O_\tau$. For the employees in the insured population these additional rights are usually funded by (possibly significant) contributions by the plan sponsors. However, the pension fund does not receive contributions to compensate for the cost due to the indexation of the rights and the payments of the deferred members and the retirees. Thus, this cost has to be met by the pension fund itself. Owing to the ageing population this cost will grow significantly, and is therefore referred to as the ghost of inflation.

In The Netherlands the discount rate is usually taken to be equal to 4 per cent, which is frequently exceeded by the total return of the investment portfolio. Thus, return on investment will frequently yield surpluses on the pension fund's balance sheet. These surpluses are used to defy deficits, to meet the cost of indexation, and to provide rebates to the contributions. The contributions minus the (possibly negative) rebates are referred to as the net contributions.

13.2.2 Asset–liability management

The contributions are paid for either by the employer, or partly by the employer and partly by the employees. Obviously two important objectives of these plan sponsors are to minimize the level of the contributions as well as the fluctuation from one year to the next. On the other hand, the pension fund by law has to pursue a third objective of minimizing the occurrence of deficits. The object of ALM is to provide an optimal trade-off between these three conflicting objectives, which can be accomplished by the determination of a coherent funding and investment policy, taking into account the development of the pension fund's liabilities.

In ALM first of all the issue of the amount of risky securities in the strategic asset mix has to be settled. Based on historical observations one may conclude that the expected return is increased if the risky securities in the portfolio are increased at the expense of fixed-income securities. Thus, additional investments in stocks and real estate will yield a reduction of the average net contributions. However, total returns of risky securities exhibit a larger standard deviation than fixed-income securities, implying larger fluctuations of the contributions and/or a larger probability of deficits. These drawbacks may be partly remedied by an appropriate funding policy, employing a minimal surplus level to meet losses of the investments and/or a maximum allowable fluctuation of the net contribution rate from one year to the next. Second, ALM has to anticipate the future costs of indexation due to inflation, which can be partly coped with by exploiting asset categories whose total returns hedge these inflation risks. For the mathematical models which are developed to clarify or solve these problems we refer to Section 13.3. Readers with no background (or interest) in the mathematical aspects of ALM may immediately turn to the self-contained Section 13.4, which contains practical experience with the application of these models to support the ALM decision process.

13.3 The OR model

13.3.1 Introduction

Figure 13.1 depicts the view of an operational researcher on ALM. That is, determine an optimal coherent investment and funding policy such that the (possibly conflicting) objectives of the level of the net contributions, the fluctuation of the net contributions and the probabilities of deficits are minimized, taking into account the development of the insured population, the sensitivity of the premium reserve to inflation, and the future returns on investments and inflations.

The approach of the ALM problem which is implemented in the decision support system is illustrated in Figure 13.2. First many scenarios are generated by economic time series of the inflation of prices and wages, the total returns of stocks, bonds and real estate, and related factors such as the growth of gross national product (cf. Section 13.3.2). Next, for each of these scenarios the corresponding actuarial time series are simulated for the salaries, contributions, payments and premium reserves of the pension fund (cf. Section 13.3.3). These combined scenarios of the economic and actuarial

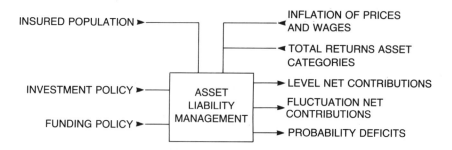

Figure 13.1 The asset–liability management problem.

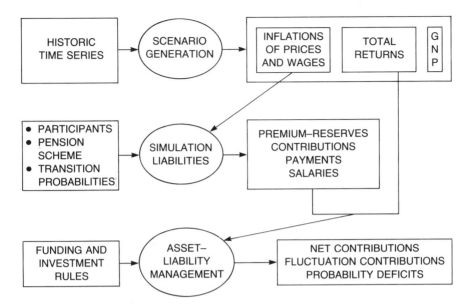

Figure 13.2   Scenario system for asset–liability management.

time series comprise the input of a simulation and an optimization model for ALM (cf. Section 13.3.4). For an integrative perspective on using scenarios to support strategic planning we refer to Bunn and Salo (1993) and the references cited therein. For a discussion on the use of scenarios consistent with statistical expectations for ALM, we refer to Kingsland (1982).

13.3.2   Generation of scenarios of economic time series

Following the strong advocation of Sims (1982) a vector autoregressive Gaussian process is used to model the economic time series. Based on historic data the order of the process is chosen to be equal to one, such that the model reads

$$\ln(1 + \mathbf{y}_t) = \mathbf{N}(\boldsymbol{\mu} + \Omega \cdot [\ln(1 + \mathbf{y}_{t-1}) - \boldsymbol{\mu}], \mathbf{C}), \tag{13.3}$$

where the notation $\mathbf{N}(\boldsymbol{\mu}, \mathbf{C})$ is used for a Gaussian distribution with mean vector $\boldsymbol{\mu}$ and covariance matrix $\mathbf{C}$, and $\mathbf{y}_t$ is the vector containing the values of the series in year t.

In the decision support system several estimation procedures can be applied to estimate the parameters of (13.3). In the first approach it is requested that $\mathbf{y}_t$ for $t \rightarrow \infty$ converges to a stationary process for which the matrices of the covariances and autocovariances are equal to the historic covariances **V** and autocovariances **W**. In Boender and Romeijn (1991) it is shown that this condition is satisfied if and only if

$$\Omega = \mathbf{V} \cdot \mathbf{W}^{-1} \tag{13.4}$$

$$\mathbf{C} = \mathbf{V} - \mathbf{W} \cdot \mathbf{V}^{-1} \cdot \mathbf{W}^{\mathrm{T}}. \tag{13.5}$$

In the second approach each row of the Ω matrix is estimated separately by ordinary least squares (OLS). It is interesting to note that in the case of infinitely many historical observations, the first approach and row-wise OLS yield identical estimates. The third approach applies seemingly unrelated regression (SUR) (cf. Judge *et al.*, 1988). That is, first the statistically insignificant parameters of Ω are removed, and the remaining elements of Ω are estimated simultaneously by generalized least squares (GLS). Note that it can be shown that OLS and SUR yield identical estimates if the matrix of the covariances of the errors is diagonal, or if each element of Ω is used as an explanatory variable in the regression.

Given the initial value of the vector $\mathbf{y}_0$, and estimated values of $\boldsymbol{\mu}$, Ω and **C**, (13.3) is applied iteratively to generate a multitude of realizations of the Gaussian process, which are used as scenarios in the subsequent ALM models.

13.3.3 Generation of scenarios of actuarial time series

For each scenario of economic time series the corresponding time series have to be computed for the development of the actuarial quantities of the pension fund (cf. the middle module in Figure 13.2). To obtain these time series the following steps are carried out. First, a push Markov model is used to determine the future development of each trustee. For an employee this implies that each year it is determined if he or she remains alive, retires, resigns, gets disabled and/or gets a promotion to a higher category, independent of the situation of all other employees. These transitions are determined by probabilities which depend on the characteristics of individuals such as age, sex and employee category. Disabled, deferred members and retirees are treated analogously.

Next, given the situation of each current employee in each future year, the pull part of the model determines additional promotions and hires new employees such that the number of employees in each category in each future year is as much as possible in accordance with prespecified values. The age and sex of new employees are random variables which depend on the categories to which they are assigned.

After the previous steps have been carried out the future situation is known for each current and future trustee. Then, taking into account the inflation of the economic time series, the pension rules (which may take up a hundred pages) are applied to compute the corresponding future salaries, contributions, payments, and premium reserves. These combined scenarios of economic and actuarial time series are input to the ALM models which are described in the next section.

13.3.4 Asset–liability management models

The decision support system contains a simulation and an optimization model for ALM. The input to the ALM models consists of vectors $\mathbf{y}_{ts}$, which denote the values of economic variables, such as inflation, and of actuarial variables, such as the premium reserve, in year t in scenario s, where the number of years and scenarios is frequently equal to 35 and 1000 respectively.

Simulation: In applying the simulation model the ALM practitioner has to specify an asset and a funding policy (cf. Figure 13.1). As for the asset policy, the following points have to be observed:

- First of all an initial portfolio has to be specified, which in The Netherlands frequently consists out of more than 70 per cent of long-term bonds and loans, and less than 30 per cent of stocks and real estate.

- Second, a choice has to be made with respect to the valuation of bonds and loans (the market value principle is applied to risky securities). The decision support system provides two alternatives, i.e. face value and net present value. Let r_{its} denote the interest rate of security i in future year t in scenario s. Then the latter approach determines the net present value of the future cash flows (interest and redemptions) of each security i, assuming a flat yield curve at the level of the current interest rate r_{its}.

- Third, an adaptation scheme of the portfolio has to be determined. Again the system provides two choices, i.e. a buy-and-hold and a rebalancing policy. In the rebalancing policy the user of the system can specify an asset mix for each future year, which will be realized at a minimum trading volume from the previous one.

For the funding policy, the following is relevant. Since the returns on the investment portfolio will frequently exceed 4 per cent, the contributions may result in large surpluses. On the other hand, owing to inflation and poor portfolio returns, the pension fund may occasionally be confronted with very low surpluses, or even deficits. To remedy these situations a minimal and maximal level of the surplus rate has to be specified (equal to the ratio of the surplus and the premium reserve). If the surplus rate exceeds the maximum allowable level, then a rebate of the contributions is determined such that the surplus rate attains its maximum level. Analogously, if the surplus rate deteriorates to a level under the agreed minimum, the plan sponsor has to donate additional contributions such that the minimum level is attained. Finally, in order to prevent the plan sponsors being confronted with too large fluctuations of the contribution rate from one year to the next, the ALM practitioner can restrict these fluctuations to a prespecified maximum level.

Given the generated combined scenarios and a specified asset and funding policy the simulation module for each scenario for each future year determines the funding level and the value of the net contributions (contributions minus rebates). These values are used to determine the average net contributions, the fluctuations of the net contributions and the probabilities of deficits referred to in Figure 13.1. Analysing these results, the ALM practitioner specifies alternative scenarios and policies until an ALM policy is achieved which provides an acceptable trade-off between the conflicting objectives of the pension fund and its sponsors. For some practical experience with this learn-and-react application of the model we refer the reader to Section 13.4.

Optimization: The optimization model is based on a funding policy which (in The Netherlands) is referred to as clean funding. That is, each year a rebate or addition to the contributions is specified such that the surplus rate is exactly equal to a fixed prespecified percentage. Given a fixed surplus rate the optimization model determines asset mixes which minimize the variances of the future net contribution rates (equal to the ratio of net contributions and salaries), given prespecified average contribution rates — referred to as q in the model.

More specifically the model reads

$$\text{minimize} \sum_{t=1}^{T} \sum_{s=1}^{S} [N_{ts}(x) - q]^2/(T \cdot S) \qquad (13.6)$$

$$x_{it} \geq 0$$

$$\text{subject to} \sum_{t=1}^{T} \sum_{s=1}^{S} N_{ts}(x)/(T \cdot S) \leq q$$

$$\sum_{i=1}^{I} x_{it} = 1 \quad (t = 1, \ldots, T),$$

where

$$N_{ts}(x) = \{(1 + e_t)R_{t-1,s} + C_{ts} - [(1 + e_{t-1})R_{t-2,s} + C_{t-1,s} - P_{t-1,s}]$$

$$\times \sum_{i=1}^{I} x_{i,t-1}(1 + r_{i,t-1,s})\}/S_{ts} \qquad (13.7)$$

$x_{it} = $ percentage of the total portfolio in asset category i in year t
    $(i = 1, \ldots, I; t = 1, \ldots, T),$

x = vector with components $x_{i,t-1}$ (i=1, . . ., I)

r_{its} = total return of asset category i in scenario s in year t

    (i = 1,. . .,I; t = 1,. . .,T; s = 1,. . .,S)

e_t = surplus rate in year t

S_{ts} = salaries in year t in scenario s

P_{ts} = pension payments in year t in scenario s

R_{ts} = premium reserve in year t in scenario s

C_{ts} = contributions primo year t in scenario s

$N_{ts}(x)$ = net contribution rate primo year t in scenario s.

Thus, analogously to the Markowitz approach to determine efficient portfolios, the optimal asset mixes can be determined by minimizing a quadratic objective function subject to linear constraints. However, contrary to the usual Markowitz portfolios, these portfolios explicitly take into account the future development of the liabilities, and they optimally balance the risks and revenues for the pension fund and its sponsors, rather than the risks and revenues of (only) portfolio return. For real-life numerical results the reader is again referred to Section 13.4.

13.4 ALM projects

13.4.1 Project team

If an ALM analysis has to be carried out, a project team is installed whose members usually include an actuary, a fund manager and an information manager of the pension fund, the actuarial counsellor, a representative of the plan sponsor (such as the controller or the manager of finance of the company), and an ALM consultant. This ALM team decides on the functionality of a supporting system, determines preferences and restrictions with respect to the objectives and instruments of ALM, and has to determine and justify an ALM policy. This policy, accompanied with the underlying assumptions and a thorough sensitivity analysis, is propounded to the board of governors of the pension fund, which ratifies the ALM proposal and takes responsibility for its consequences.

13.4.2 Acquisition

Clearly an important objective of our company is to be selected as the consultant in these ALM teams. In The Netherlands almost 10 competitors are active in the battlefield of decision support systems for ALM. In contrast to our company, the ALM competitors are all subunits of large companies such as actuarial counsellors, insurance companies and fund managers. Nevertheless, once we established contact with a potential new client, we (until now) never lost it to one of the competitors, resulting in a current clientele of pension funds with an investment portfolio of more than 50 billion Dutch guilders, and more than 1 million trustees. In this section we will focus on success factors of acquisition, and illustrate these with our ALM experiences.

Expertise: To carry out a successful ALM project several disciplines have to be mastered:

- Econometrics and statistics to model the economic time series.
- Labour planning to determine realistic scenarios of the future development of the employees and other trustees.
- Actuarial sciences for the pension schemes and the funding policy.
- Finance for asset allocation.
- Operational research for modelling the ALM problem, and for simulation and/or optimization of ALM strategies.
- Computer science for building an efficient, user-friendly and maintainable tailored software system.
- Consultancy and decision support to make the system really useful for the decision makers.
- ALM which embraces all previous disciplines to determine a balanced asset and funding policy, taking into account all relevant expectations and uncertainties.

In an acquisition stage each of these disciplines may turn out to be crucial. The actuaries of a pension fund are not only capable of judging the required actuarial background of the consultant, but they usually can evaluate the extremely important expertise with respect to the modelling and estimating of the economic time series as well. Second, in an acquisition stage, the consultant has to convince the fund managers of the ALM team, which evidently requires a sufficient background in asset allocation and asset evaluation models. Finally, and worst of all, there is the problem of convincing the director(s) responsible for information, and software and hardware, which requires a vision and expertise with respect to object-oriented programming, system development methodology, MS-DOS(-extenders), UNIX, WINDOWS(-NT), downloading, relational data bases, interfaces and other vocabulary that the author of this chapter is only reluctantly familiar with.

Generally speaking, each potential client for each relevant aspect of the problem simply demands the 'best' from the consultancy office. Any flaw which is detected by the potential client in the acquisition stage frequently will turn out to be fatal. This required expertise concerns the theory of OR, familiarity with the problem environment, the software, etc., which evidently makes applied OR a challenging but hard profession.

Reputation: Unfortunately, and for many operational researchers also surprisingly, expertise is necessary, but certainly not sufficient to acquire new clients. In addition OR consultancy offices increasingly need a reputation in the field of the problem which a potential client wants to be clarified or solved. In ALM our reputation has been accomplished by studying the literature, carrying out and presenting fundamental research, implementing the results in the decision support systems, contributing to seminars which are attended by the managers of pension funds, and nourishing our clientele. As remarked in Section 13.1.2 these activities were initially carried out largely at our own cost after completing our first project for a pension fund. Later, one after the other, new clients and co-investors joined in, and each new acquisition enhanced the probability of acquiring the next.

In our opinion it is a general tendency that references and demonstrable experience are increasingly valuable assets in acquiring profitable new OR projects, and that the acquisition of new clients by an OR company concerning a problem which it has not tackled so far is becoming prohibitively difficult. This implies that OR companies increasingly have to specialize on certain problem areas, analogously to universities which for similar reasons have to concentrate on major research themes.

Attitude: A third and frequently deplorably misjudged and underestimated success factor in acquisition is the attitude of the OR consultant. Far too often, the grasp and the interest of operational researchers is too much confined to models, algorithms and software, whereas the one and only focus in acquisition should be the client and his or her problem. The introduction of an OR system almost always touches other parts and persons of the client's organization, and meets with several difficulties, for which the OR consultant should have a skilful eye. A vehicle routing system requires an automated processing of order data, and threatens the responsibility, freedom of movement or even the employment of the planners and drivers. The use of an ALM system brings to the surface the possibly conflicting objectives of the ALM decision makers, and several other important and delicate issues such as the assumptions to evaluate the liabilities and the asset categories. The ability to think beyond the OR characteristics of a problem, and the attitude to work problem oriented, rather than discipline oriented, is often crucial for the client in allowing the consultant the job.

13.4.3 Decision support system

Once we are included in an ALM team at a pension fund, the project takes off by setting a number of important initial points of departure, such as:

1 The specification of possible ALM policies which will be analyzed, i.e.:
   ■ the choice of strategic asset mixes and rebalancing policies;

- the choice of funding systems, including a minimal and maximal funding level, and a maximal fluctuation of the contribution rate from one year to the next.

2 The specification of scenarios of basic assumptions to determine the future consequences of ALM policies, such as:

- the economic time series which will be used;
- the historic periods which are used to generate scenarios of future time series;
- the estimation procedure to specify the model (13.3);
- the method to appreciate the several investment categories.

3 The specification of the ways in which the results will be presented, both numerically and graphically.

4 Preliminary quantification of preferences and restrictions with respect to the conflicting objectives of ALM.

Once these initial choices have been made, the consequences of the possible ALM policies have to be computed, analysed and evaluated, which (fortunately) needs the development of a tailored decision support system.

Our point of departure to develop such a tailored ALM system is a system containing the models and methods which are described in Section 13.3. The system is programmed in C (object oriented), and runs under MS-DOS and UNIX, applying a DOS extender to cross the boundary 640 Kb memory. To simulate the consequences of an ALM policy for a pension fund of about 50 000 trustees, based on 1000 scenarios over a time period of 35 years, the system requires less than one hour computation time on an 80486 66 MHz processor. The computation of an optimal time-dependent asset allocation described in Section 13.3.4 roughly takes the same amount of time.

Starting from the basic system, the first adjustment which has to be carried out is the implementation of the current pension rules of a pension fund, and the possible conversions which are considered. These conversions not only concern the transition from final pay systems to average pay systems, but also the introduction of partial, early and flexible pension schemes, and the extent to which the earned rights of the trustees are compensated for inflation. These pension schemes differ enormously in size and variety and take an amount of work (significantly decreasing over time) varying from about a week to 100 days to implement.

Until now, fortunately, it has never occurred that the tailoring activities remained confined to the pension scheme. It always appeared necessary to specify and implement additional extensions and improvements which fit more in the working area of the OR consultant. Some memorable examples include:

- The inclusion of the dynamic funding system (for and with our first client and others).

- An alternative for the modelling of careers, in which the increase of an employee's salary, excluding inflation, is determined as a function of age (for and with the pension fund of a large oil company).

- Presentation of the consequences of an ALM policy by the balance sheets and profit and loss accounts (for and with the pension fund of a large bank).

- Flexible compensations of the earned rights of trustees for inflation (for and with the pension fund of a large industrial concern).

- Improvement of the estimation procedure of the time-series model (13.3) (for and with the research institute of a large fund manager).

- Extension of the simulation model for ALM, including the design and implementation of important graphical presentations of results (for and with an insurance company).

Almost always, the extensions and improvements which are requested by a new client turn out to be generally applicable, and with permission are put at the disposal of existing clients, such that not only new clients profit from the previous ones, but also the other way around.

The required data of an ALM system consist of the information of all trustees, the transition probabilities, the historic realizations of the economic time series, and scenario data. Although time consuming, the gathering of these data is less cumbersome than in most other applications of OR. This is due to the fact that most of these data are required by the pension fund to carry out the daily activities of determining earned rights, determining and paying of payments, taking investment decisions, etc. An exception to this availability of required data is the function category or salary group of employees. These data are necessary for the career model described in Section 13.3.3, but are frequently not at the disposal of the pension fund. For cases where these data are unobtainable we use the simple career model described above. Testing of the system is usually carried out specifying a representative set of, say, 100 trustees, and comparing the actuarial quantities of these trustees which are computed by the system, with hand-made computations.

13.4.4 Computational results

Once a tailored version of the system is completed, the ALM team starts the process of decision making. Before we turn to the parts which are played by the OR consultant and the decision support system in this process, we first present some results which are computed by a version of the system. The results concern a small pension fund of about 3000 trustees, consisting of 2000 employees, 1000 resigned employees, 100 disabled and 150 retirees. In 1991 the salaries, earned rights, premium reserve and payments are equal to 64,

31, 156 and 4.5 million Dutch guilders respectively. The pension fund carries out an average-pay scheme, where the employees build up old-age pension rights of 1.75 per cent of their salary per year until retirement at the age of 65. Widow(er)'s pension is equal to 70 per cent of old-age pension, and all earned rights are compensated each year for price inflation, excluding the rights of the employees which increase with the inflation of wages. The discount rate is equal to 4 per cent.

To apply the simulation and optimization models of Section 13.3.4 we generated 1000 scenarios of (Dutch) wage and price inflation, growth of gross national product, and the total returns of cash, long-term bonds, stocks and real estate over the period 1992–2026, using the OLS method and the realizations over the period 1952–91 to estimate the time series model (13.3). Some relevant statistics of these economic time series are contained in Table 13.1.

Next, for each scenario we simulated the corresponding future development of the salaries, payments and premium reserves. These simulations are carried out using transition probabilities of survival, career, resignation and disablement which are estimated using real-life data over a period of five years. The acquisition of new employees is carried out such that the number of employees in each salary group remains in accordance with the situation in 1991. Given these 1000 combined scenarios of economic and actuarial time series we simulated four ALM policies using the simulation model of Section 13.3.4, and computed an efficient frontier of optimal asset mixes using the optimization model of Section 13.3.4.

Simulation: We computed the results of four ALM policies whose definition and results are depicted in Figure 13.2. Each value in the table is an average, standard deviation or probability which is based on 35 000 observations (1000 scenarios over a period of 35 years). Note that the volatility is defined as the volatility of the contribution rate from one year to the next, whereas the standard deviation measures the spread of the contribution within one year.

From Table 13.2 the reader can verify the impact of changes in the strategic asset mix and the funding policy. Of great importance is the sensitivity analysis with respect to the historic period, which should convince ALM decision makers to be extremely careful in the interpretation and appraisal of the results.

Optimization: Next, we computed an efficient frontier of asset mixes which for a given average value of the contribution rates yields a minimal variance. Note that these optimal asset mixes are based on the same set of 1000 combined scenarios of economic and actuarial time series as the simulated policies in Table 13.2.

It is interesting to analyse the part which is played by real estate. From Table 13.1 it can be observed that the returns of real estate have a lower

Table 13.1 Statistics based on the period 1952–91

	Wages	Prices	Cash	Bonds	Stocks	Real-estate	GNP
Average value, μ	6.4%	3.9%	5.7%	5.5%	10.0%	5.3%	3.6%
1991 value, y_0	3.8%	3.9%	9.6%	12.4%	11.2%	4.0%	2.2%
Standard deviation	4.1%	2.7%	2.8%	5.1%	15.5%	4.6%	2.4%
Autocorrelation	0.52	0.83	0.75	0.14	−0.01	0.68	0.45
Correlation with wages	–	0.68	−0.02	−0.15	−0.21	0.58	0.25
Correlation with prices	0.68	–	0.37	0.20	−0.08	0.52	−0.06

Table 13.2 Results for the four ALM policies

Bonds and cash	Risky assets	Surplus Min	Max	Historic period	Return on investment		Prob. of 1 year	Deficits ≥3 years	Contribution rate level		Volatility
70%	30%	15%	0%	1952	7.0%	(4.8%)	17.9%	5.2%	13.4%	(10.5%)	6.9%
100%	0%	15%	0%	1952	6.0%	(4.2%)	30.8%	13.5%	15.9%	(9.6%)	6.3%
70%	30%	5%	5%	1952	7.0%	(4.8%)	13.9%	0.4%	13.4%	(15.8%)	15.2%
70%	30%	15%	0%	1972	8.1%	(5.0%)	11.4%	3.7%	9.2%	(13.2%)	7.8%

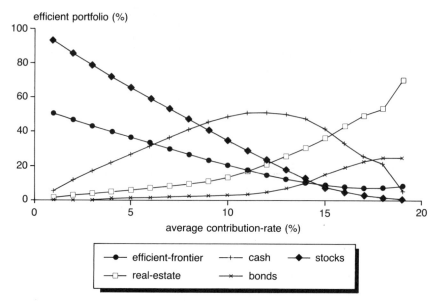

Figure 13.3 Efficient frontier and efficient portfolios with fixed surplus rate of 5 per cent.

average value and a larger standard deviation than the returns of cash. Hence, if only return is considered, real estate is inefficient with respect to cash. Nevertheless, real estate appears in significant amounts in the efficient portfolios in Figure 13.3. This is clearly the result of the positive correlations of 0.52 and 0.58 on the return of real estate with the inflation of prices and wages. Thus, although the real estate in this chapter (value increase of Dutch homes) is not representative of the investments in real estate of many pension funds, these outcomes nicely demonstrate the quintessence of the optimization model.

13.4.5 Decision support

Once the tailored system is completed an ALM project team usually meets more than 10 times before a final ALM policy is established and agreed. In this chapter we will not discuss which ALM policies the project teams which we joined have agreed upon. Neither will we use the results like the ones described in Section 13.4.4 to discuss which ALM policies in our view and experience optimally fit pension funds with a certain size, maturity and pension scheme. These conclusions will be presented in a paper to be submitted to a financial journal. In this section we will rather focus on the decision-making process in the ALM teams, with special emphasis on the role of the operational researcher and the decision support system. In our experience this role primarily boils down to enhancing the understanding of ALM complexities by the team members, contributing to reaching consensus between the team members, and, if necessary, the specification and implementation of extensions which appear necessary to endorse an ALM policy.

With respect to extensions which appeared necessary in a very late phase of an ALM project, we confine ourselves to one example, which is the inappropriateness of the assumed lognormal distribution of the inflation of wages. It appeared that scenarios of the time series of wage inflation were generated which badly fitted practical observations. The reason is that due to the assumed lognormal distribution too much low wage inflation is generated, which obviously results in an underestimation of the future contributions. This problem has been solved by determining parameters α and β such that the logarithmic transformation $\ln(\alpha + \beta \times$ wage inflation) maximally fits the historic observations (rather than $\alpha = \beta = 1$).

With respect to the improvement of the understanding of ALM complexities, it is crucial that the OR consultant keeps explaining the assumptions and the degree of realization of the underlying models to the project team. To accomplish this, we replicated the so-called 80/20 rule from logistics. That is, the level of the future probabilities of deficits and contribution rates should be assigned a reality rate of 20 per cent, but the difference between the values of these quantities resulting from different sets of assumptions should be assigned a reality rate of 80 per cent, where '80' and '20' should be interpreted as figurative numbers. To emphasize this view, we always insist that the results of an ALM policy are computed for different historic periods, as illustrated in Table 13.2. We also refer to the references of Kaplan (1993), Michaud (1989) and Lummer et al. (1993) for the sensitivity of the results of mean–variance optimization to statistically formed inputs, and the required sound understanding of capital markets for a proper use of these techniques. Consistent with this view is the (encouraged) possibility to 'overrule' the assumptions of the model, by the own perceptions of the decision maker with respect to future developments. The most relevant example is the substitution of the long-term averages of the scenarios of economic time series generated by model (13.3) by the decision maker's own expectations.

The usefulness of decision support systems to facilitate the accomplishment of consensus is well known. In ALM each of the issues 1 up to 4 described in Section 13.4.3 always gives rise to long discussions, in which the preferences and responsibilities of each of the members of the team quickly come to the surface. A first delicate issue is the choice of the strategic asset mixes which will be analysed. Usually the fund managers of the pension fund and the plan sponsor have a preference for a relatively large amount of risky securities in the strategic asset mix, whereas the actuaries frequently opt for a lower-risk profile. A second issue is the choice of the minimal required funding rate, which specifies which of the two parties (pension fund or plan sponsor) is responsible for the elimination of underfundings. However, several other choices have a large impact on the risk/return issue as well, such as the method of evaluating bonds. Clearly, if the value of a bond is determined as the net present value of the future cash flows, rather than its face value, much larger deviations of portfolio return will result, and hence much larger probabilities of deficit. In practice it appears that the use of a decision support system makes it almost impossible for team members to hold on to prejudiced or traditional, but poorly founded, opinions about the above issues. The reason is that the system confronts the team members with the numerical consequences, which frequently incite them to reconsider and revise their views. This contribution of the system once elicited one project member to make the observation that the system is not only decision supportive, but opinion supportive as well.

Finally, it could be argued that the application of an ALM system can be 'profitable' since it can endorse the decision to increase the investments in high-return risky securities, and in securities which hedge the pension fund against the risk of inflation. However, we never use this as the strongest selling point of the system. In our opinion more value should be assigned to the application of the system to get to know the pension fund, and how it interacts with economic reality, such that preferences and opinions can be stated explicitly, and better decisions be made in consequence.

13.5 Conclusions

13.5.1 Future ALM business and research

Our optimization model for ALM can be viewed as a pension fund adjustment of the Markowitz mean–variance model for asset allocation. The adjustments are that our model is multi-period, it uses scenarios, it takes into account the liabilities, and the trade-off of risk return is made with respect to the objectives of an asset–liability manager, rather than a portfolio manager. Nevertheless this model can be significantly improved. At the top of the list is the extension such that a minimum and maximum funding level can be handled (rather than a fixed funding level), and such that probabilities of underfunding can explicitly be taken into account. These extensions have not been carried out so far since they appear to ruin the quadratic programming formulation of the optimization problem. One of the possible approaches for this extended model is the scenario aggregation method of Wetts and Rockafellar (unpublished manuscript). A second extension is the incorporation of the possibility of computing adapting policies in which the ALM policies not only depend on the year t, but also on the scenario s. This problem can be formulated as a multi-stage stochastic programming problem, which is the subject of the PhD thesis of C. L. Dert at the Erasmus University Rotterdam.

With respect to the other models only a few extensions are currently considered to be expedient: the possibility of generating scenarios of time series which exhibit (abruptly) changing trends, the incorporation of expert knowledge of investment managers in the ALM simulation model, and a more sophisticated procedure to handle underfundings and overfundings.

Other, and quite different, actions which are considered are to use the current references, software and expertise to conquer Europe, and/or to cope with the (different) ALM issues of insurance companies and banks.

13.5.2 Propositions and recommendations

- Successful applied OR is impossible without a thorough and broad theoretical background.
- Applied operational researchers should work problem oriented, understand the problem they are modelling, and have a skilful eye for the consequences of the implementation of OR models in the client's organization.
- Applied operational researchers should develop a spirit of enterprise, create a network of connections, be willing to make (risky) investments in know-how, software development and publicity, and able to make these investments a business success.
- Universities should attract not only the most intelligent students, but also clever, ambitious ones, not give in to reductions of quantity and quality of education, and focus on main research themes in which external OR practitioners and problem experts also participate.
- OR consultancy offices should attract not only the most intelligent postgraduate students, but also clever, ambitious ones, invest in know-how, software development and publicity, and focus on main application areas, taking advantage of the knowledge and expertise of university colleagues and clients.

13.5.3 Lessons to be learned

It is our experience that the key factors of successful applied OR projects are very adequately grasped by the 'lessons' of the late M. Beale (Hughes, 1990):

- Understand the environment that you are modelling.
- Produce some results quickly to whet the client's appetite and to clarify the really important issues.
- Do not get slavishly tied up to a software package.
- Always keep sight of the numbers to ensure that the problem is correctly formulated.

In our opinion additional lessons for successfully applied OR are:

- Do not immediately degrade to problem solving, but take advantage of the theory of OR.
- Do not use models and methods to a level of sophistication beyond the grasp of your client.
- Work problem oriented, rather than discipline oriented, and be a creative consultant, rather than a slavish technocrat.

In applying these lessons ORTEC does not make a sharp distinction between marketeers, operational researchers, and computer programmers. Everybody, including the computer programmers, understands the environment of the projects, and is familiar with the basics of OR. Also, being problem oriented, the company focuses on the development of tailor-made systems, rather than on selling commercial software packages. Finally, in attempting to solve problems as efficiently and effectively as possible, we frequently call on experts at several universities (to the benefit of all).

Acknowledgements

We acknowledge (in alphabetical order) the valuable contribution of many of our colleagues and clients at AMEV-FORTIS, Coopers and Lybrand, Erasmus University Rotterdam, IRIS, FIST, ORTEC, PIRI, Stichting Pensioenfonds RABObankorganisatie (and others).

Appendix: the consultancy bureau

The core business of ORTEC Consultants bv is the development, implementation and application of major decision support systems, containing models and methods from the disciplines of operations research, management science, and finance. The company was founded on 1 April 1981 by operations researchers who received their masters degree from the Econometric Institute of the Erasmus University Rotterdam. The company now employs about 40 people whose backgrounds are in operations research, econometrics, mathematics, finance and computer science.

The company is active in many of the areas of operations research applications, with special emphases on strategic planning, transportation, production planning and scheduling, crew planning and scheduling, risk analysis, porfolio management, and asset liability management. Projects include:

- An interactive optimization system which is used by a sugar refinery with several depots for routeing and scheduling of its bulk vehicle fleet.
- A hierarchical multi-criteria decision support system, which has been

developed in cooperation with the research and development department of the Dutch Government Buildings Agency to optimize the allocation of the available office buildings and offices to the organizations and (more than 150,000) employees which have to be accommodated.

- Consultancy of a national industry with tens of independent businesses with respect to strategic issues, such as buying-out of businesses, and the employment of joint trucks to reduce cost and/or improve in-time deliveries.

- A joint project of the Erasmus University Rotterdam, an international dredging company and ORTEC, resulting in a PhD of one of the participants, and a decision support system for risk analysis of investments and tenders.

- Tailor-made decision support systems and consultancy for asset liability management of several pension funds.

References

BOENDER, C. G. E. and ROMEIJN, H. E. (1991) The multi-dimensional Markov chain with prespecified means and (auto-)covariances, *Communications in Statistics: Theory and Methods*, **20**, 345–60.

BUNN, D. W. and SALO, A. H. (1993) Forecasting with scenarios, *European Journal of Operational Research*, **68**, 291–301.

HUGHES, H. (1990), Martin Beale: a personal memory, *Mathematical Programming*, **42**, 1.

JUDGE, G. G., HILL, R. C., GRIFFITHS, W. E., LUTKEPOHL, H. and LEE, T. (1988) *Introduction to the Theory and Practice of Econometrics*, New York: John Wiley.

KAPLAN, P. D. (1993) Asset allocation models using the Markowitz approach, *OR/MS Today*, (April), 18–23.

KINGSLAND, L. A. (1982) Combining financial and actuarial risk: simulation analysis, *The Journal of Finance*, **37** (2) 577–94.

LUMMER, S. L., RIEPE, M. W. and SIEGEL, L. B. (1993) Taming your optimizer: a guide through the pitfalls of mean variance optimization, in: J. B. Lederman and R. A. Klein (eds), *Advances in Asset Allocation Techniques for Optimizing Portfolio Management in the U.S. and Global Markets*, New York: John Wiley.

MICHAUD, R. O. (1989) The Markowitz enigma: is 'optimized' optimal?, *Financial Analysts Journal*, (January/February), 129–137.

SIMS, C. A. (1982) Macroeconomics and reality, *Econometrica*, **48**, 1–48.

Bibliography

BOENDER, C. G. E. (1989) Kwantificeren en analyseren van projectrisico's, *Het Ingenieursblad*, **58**, 7–8.

(1995) A static scenario optimization model for asset liability management of defined benefit plans, Erasmus University Rotterdam, Econometrical Institute, Report 9512/A.

BOENDER, C. G. E., VAN HETTEMA, I., JANSZEN, I., PULLEN, W. R., STEGEMAN, H., TAS, A. and WASSENAAR, C. L. G. (1993) A decision support system for housing of (public) organizations, in: Timmermans, H. (ed.) *Design and decision support systems in architecture*, Dordrecht: Kluwer Academic Publishers.

VAN VLIET, A., BOENDER, C. G. E. and RINNOOY KAN, A. H. G. (1992) Interactive optimization of bulk sugar deliveries, *Interfaces*, **22** (3) 4–14.

VELLEKOOP, A. H. (1990) Decision support for the analysis of economic risk, PhD Thesis, Erasmus University Rotterdam.

CHAPTER FOURTEEN

A chain model for potted plants

E. H. POOT and E. M. T. HENDRIX

14.1 The environment

The Netherlands is among the leading countries in the production and distribution of ornamental products (cut flowers, indoor plants and outdoor plants). In 1992 the production value of Dutch ornamental products was more than 5 billion Dutch guilders. The number of cut flower transactions was 9.6 billion, the number of indoor plants (or potted plants) 0.6 billion and the number of outdoor plants 0.3 billion. The product differentiation is enormous: there are about 5500 cut flower varieties, 2000 potted plant varieties and 2000 outdoor plant varieties. Ornamental products are classified on length (roses), weight (chrysanthemums), diameter, height (potted plants) and quality. Despite these possibilities for classification, it turns out to be hard to describe products objectively. This lack of standardization obstructs the introduction of efficient logistical systems. Another problem in ornamental logistics is caused by the perishable character of the products. About 15 to 20 percent of the products never reach the consumer.

A central place in ornamental distribution is taken by flower auctions. At the flower auctions, the price of the products is formed by physical concentration of supply and demand. At the auctions there are many suppliers (about 7000) and many buyers (about 2000). Together they constitute a complex network of distribution chains (Figure 14.1). This network is of great commercial importance, but also puts high demands on the logistical management.

Flower auctions play an important role in managing ornamental logistics, because almost every product physically passes the auction. The number of transactions grows faster than the turnover, so logistical costs are expanding. Therefore the demand for management support increases.

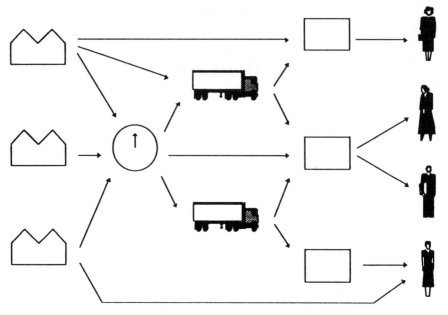

Figure 14.1   Complex network.

14.2   **The problem**

A distribution chain is one of the paths in the distribution network from primary production to the final consumer. In the potted plants distribution chain this path can go from grower, via auction clock, exporter, importer and wholesaler, to the shop holder. These so-called links in the chain are responsible for a particular function and perform logistical processes on the product to meet the requirements of the final consumer, but also for demands on the product by other links in the chain (Bowersox *et al.*, 1986). To satisfy demand, handling has to take place. For example, if the retailer demands a label, some link in the chain has to label the potted plant. The shop holder can label the potted plant, but so can the wholesaler or the grower. At first, the location of the handling process is important, because one link can attach or remove material at a lower price than another one, e.g. because a handling process is automated. Second, distribution costs are influenced by the sequence of attaching and removing materials. When, for example, the label is added to the plant by the exporter, it may be that sealing material first has to be removed and then added again to the plant for protection during transport. The number of activities can differ between handling sequences. Generally, more handling means higher distribution costs. Owing to lack of information and coordination in the chain those handling processes may not be well tuned, which causes substantial losses.

Table 14.1 Stages in potted plant distribution

Stage	Function	Link
Production	Supply	Grower
Auction	Price forming	Auction clock
		Negotiation
Wholesale	Distribution	Exporter
		Importer
		Wholesaler
Retail	Sale	Shop holder
		Garden centre
		Supermarket
		Market vendor

The flower auctions play a central role in the distribution process. Despite the fact that they do not form a coordinating institute over all the distribution network, they want to gain more insight into the distribution costs of potted plants. What amount of money can potentially be gained by better tuning the processes in a distribution chain by having the correct links doing the correct handling and by using the correct (uniform) and cheapest packing material in a distribution chain? In cooperation with the Association of Dutch Flower Auctions (VBN) the section OR of the Department of Mathematics, Agricultural University Wageningen, analysed the relations between logistical processes within distribution chains and distribution costs. We will now describe the logistical chain and the logistical processes.

During the distribution, a potted plant can pass the stages summarized in Table 14.1 (Van der Laan *et al.*, 1988).

The links cause logistical processes which can be divided into handling processes on one side and storage and transportation processes on the other. Storage and transportation processes take place when during a relatively long period the potted plant remains in the same state with respect to the materials around the potted plant. These types of processes take place when, say, a potted plant is exposed for sale or in transportation. Handling processes take place when during a relatively short period the state of the potted plant is changed. During the handling processes, materials which have one or more functions in potted plant distribution are attached or removed. Table 14.2 sums up the handling processes under consideration (Van der Laan *et al.*, 1988).

The state of a potted plant is defined by the presence of (a combination of) materials at a certain point of time during distribution. Handling processes can change the state of a potted plant. During storage or transportation, the state of a potted plant does not change.

Table 14.2 Handling processes in potted plant distribution

Handling process	Material	Function
Labelling	Label	Presentation
Wrapping	Wrapping plastic	Protection
	Wrapping paper	Presentation
Packing	Box	Grouping
	Tray	Protection
Loading on container	Container	Internal transportation
	Pallet	Grouping
Sealing	Foil	Protection
Loading into conveyance	Truck	External transportation
		Grouping

Several links in the distribution chain put demands upon the state of the potted plant during storage and transportation. These demands can be divided into three types. In the first place, market demands can be distinguished, e.g. the presence of labels is required when potted plants are exposed for sale. The second type of demand concerns quality, e.g. potted plants must be wrapped in plastic to be protected against mechanical damage and physiological deterioration during distribution. The third type of demand is logistical, e.g. potted plants have to be loaded onto trucks for long-distance transportation. In the next section we describe how the relations between the distribution chain, the state of a potted plant, handling materials and costs can be modelled.

14.3 The OR approach

The problem may be formulated as an assignment problem: assign handling activities to the links of the distribution chain, in such a way that requirements about the state of the potted plant are satisfied and the distribution costs are minimal. This forms the core of the model. The relations between the packing materials and the differing sizes of the material complicate the model further.

14.3.1 Allocation model

The assignment problem may be formulated as follows. Index the links of the distribution chain by $s = 1, \ldots, S$. Index the materials by $i = 1, \ldots, I$. This classical allocation model is now extended with an additional index 'point of time', $j = 1, \ldots, J$, to determine an explicit handling sequence.

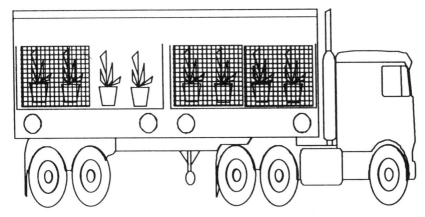

Figure 14.2 Hierarchical relations between materials.

Because of the possibility of performing more than one handling activity in a single link, several points of time per link are distinguished. We will call these points of time 'decision points', because it has to be decided if a handling activity is performed at this point of time or not. The relation between a link and decision point is described by $s(j)$: decision point j is in link s. The state of a potted plant at decision point j with respect to material i can be represented by the binary variable $X_{is(j)}$ or X_{ij}. This variable reflects the presence ($X_{ij} = 1$) or the absence ($X_{ij} = 0$) of material i at decision point j.

Let p be the point where the potted plant has to be in a certain state. The requirements made upon the state of a potted plant can be modelled by fixing the value of variable X_{ip} at 1 (presence of material i at point p is required) or 0 (absence of material i is required). The state of a potted plant can be changed by performing handling activities. Two handling activities can be distinguished. At first, binary variable Y_{ij}^+ represents attaching ($Y_{ij}^+ = 1$) or not attaching ($Y_{ij}^+ = 0$) material i to the potted plant at decision point j. Second, binary variable Y_{ij}^- represents removing ($Y_{ij}^- = 1$) or not removing ($Y_{ij}^- = 0$) material i from the potted plant at decision point j. The costs of performing handling activities depend on the links where the handling activities take place. These costs are represented by $c_{is(j)}^+$ for attaching material i and $c_{is(j)}^-$ for removing material i at decision point j in link s. This part forms the core of the model.

The application of activities is governed by so-called logical relations. When a certain state occurs, some activities cannot be performed; first this state has to be changed by performing other (removing) activities. For example, in Figure 14.2, when a potted plant has been loaded into a truck, it is impossible to pack the potted plant in a box before the truck is unloaded. Another example is when a potted plant is packed in a cardboard box, it has to be taken out of this box before it can be packed in a plastic tray. In general,

Table 14.3 Hierarchical levels of materials

Material	Material index i	Hierarchical level $h(i)$
Label	1	1
Wrapping plastic	2	2
Wrapping paper	3	2
Box	4	3
Tray	5	3
Container	6	4
Pallet	7	4
Foil	8	5
Truck	9	6

when a certain material contains a potted plant, it is not possible to perform a handling activity with a smaller material or a material of the same size. These if–then relations can be included in a binary programming model. To describe this relationship between materials, index h is introduced and represents the hierarchical level of material i: $h(i)$. Table 14.3 gives the hierarchical levels of the materials under consideration.

The optimal assignment of handling activities to the links of a distribution chain can be found by using model 1 in Appendix A. This binary programming model can be used to calculate optimal distribution scenarios. By comparing these scenarios with real-life scenarios, tuning losses can be traced. Model 1 includes the allocation idea, the sequence aspect and the relations between packing materials. While discussing the first results with VBN it became clear that VBN also wanted to evaluate the choice of alternative types of packing material. This is one aspect which the auctions can influence by promoting a type of (standardized) material. A complicating factor is that this choice of material influences the handling costs per plant. This effect will be called the load effect. Its inclusion in the model is discussed in the next section.

14.3.2 Load effect

Loading a box containing five plants implies about the same handling costs as loading a box with four plants. The costs per plant, however, will differ. This effect becomes more complex when a combination of materials on various levels is used. The load effect is the effect on the costs of one potted plant caused by the number of potted plants contained by a combination of materials. A certain truck can contain different types of containers, different types of packing material and different types of potted plants. Different

Table 14.4 Enumeration of possible packing–container combinations

k	Presence of material in combination k $h = 3$ $i = 4$	$h = 3$ $i = 5$	$h = 4$ $i = 6$	$h = 4$ $i = 7$	Number of potted plants	Costs per plant of loading a truck
1	0	0	0	0	1	100
2	1	0	1	0	10	10
3	0	1	1	0	8	12.5
4	1	0	0	1	15	6.67
5	0	1	0	1	12	8.33
etc.	. . .	. . .	. . .	. . .	. . .	. . .

combinations can contain different numbers of potted plants, which are distributed at different costs. The number of potted plants contained by the truck is restricted by:

- the dimensions of the truck;
- the dimensions of the container;
- the dimensions of the packing material;
- the size of the potted plant.

In the next example two packing materials and two internal transport materials are considered (Figure 14.3). The first packing ($i = 4$), a cardboard box, can contain five potted plants of a given size. The second packing ($i = 5$), a plastic tray, can contain four potted plants. The two packing materials have more or less the same dimensions. On a container ($i = 6$) two packings can be placed. On a pallet ($i = 7$) three packings can be placed. There are different possible combinations. Table 14.4 enumerates the possible combinations which are indexed with $k = 1, \ldots, K$. The presence of material i in combination k is represented by a one, the absence by a zero. We assume that the costs of loading material onto a truck are independent of the material combination that is loaded (but dependent on the distance between material and truck, the presence of dock shelters, labour costs, etc.). In the last column of Table 14.4, the costs of loading material (100) are divided by the number of potted plants contained by the combination of materials.

There is a load effect not only on handling costs, but also on material costs and costs of transportation and storage. These effects are calculated in the same way.

Now we can formulate a model for calculating optimal distribution scenarios, taking load effects into account. We index the enumerated material combinations by $k = 1, \ldots, K$. Let κ_k be the index set of all materials i present in material combination k. The hierarchical level of κ_k is determined by the element with the highest hierarchical level. For example, in Table 14.4

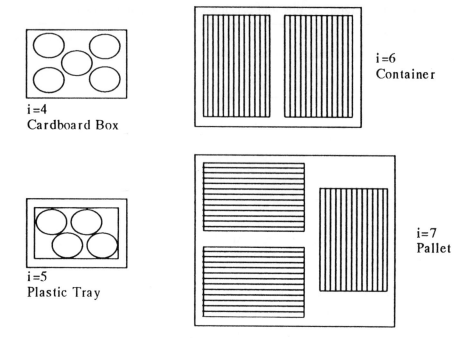

Figure 14.3   Load effect.

κ_2 is a combination of cardboard boxes ($i = 4$) and a container ($i = 6$). The hierarchical level of κ_2 is determined by the hierarchical level of the container.

To describe handling with material i at decision point j on a combination of materials k, the binary variables Z^+_{ijk} and Z^-_{ijk} are introduced. Variable Z^+_{ijk} presents attaching ($Z^+_{ijk} = 1$) or not attaching ($Z^+_{ijk} = 0$) material i to material combination k, at costs $c^+_{is(j)k}$. Variable Z^-_{ijk} presents removing ($Z^-_{ijk} = 1$) or not removing ($Z^-_{ijk} = 0$) material i, so that material combination k remains, at costs $c^-_{is(j)k}$. The result is a hugh model which can be found in the appendix (model 2). Further details can be found in Poot and Hendrix (1994).

Model 2 minimizes the distribution costs in deciding which handling activities have to take place where, in what sequence and using what type of material in which combination. In the model every activity can take place at every point of time and there is the restriction (A14.10) that at most one activity is selected. This makes the structure of the model relatively simple, but of an enormous size.

14.3.3   Reducing the size of the model

The size of model 2 can be reduced by using the particular structure of the problem resulting in a model which is less clear, but much smaller. The

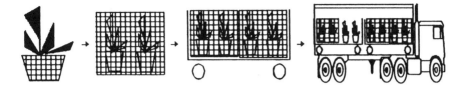

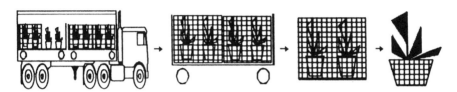

Figure 14.4 Fixed sequence.

number of variables and equations in model 2 is in general very large, because each combination of material i and decision point j is allowed. It is possible to reduce the number of variables and equations taking advantage of the fact that handling processes are performed in fixed sequences, because of the (hierarchical) relations between materials. An example of such a sequence is shown in Figure 14.4. Because of the fixed sequence of handling activities, it is possible to assign handling activities to variables before the model is solved. The variables X_{ij}, Y_{ij}^{+}, Y_{ij}^{-}, Z_{ijk}^{+} and Z_{ijk}^{-} and the balance equations (A14.9) only have to be included for specific decision points in the distribution chain. Moreover, the restriction that only one activity per period is allowed (A14.10) becomes redundant. Packing the potted plant in a box ($i = 4$, see Table 14.3) can be assigned to $j = 1$, loading on a container ($i = 6$) can be assigned to $j = 2$, etc. Variables $X_{4,1}$ and $Y_{6,2}^{+}$ are included in the model, variables $X_{4,2}$, $Y_{4,2}^{+}$ and $Y_{6,2}^{-}$ are not.

The assignment of handling to decision points reduces the dimensions of the problem considerably. Still the binary programming model 2 remains very large. We developed a heuristic method to determine good but not necessarily optimal distribution scenarios using the smaller model 1 iteratively, bringing information about load effects gradually into the problem. This will be described in the next section.

14.3.4 Local search heuristic

The binary programming model 2 decides on what, where, and using which materials in which combinations. The combination of materials giving the

Table 14.5 Decision scheme

		Retailer		Expected value
		Wrapped	Not wrapped	
Grower:	Do wrap	A	B	$pA + (1\text{-}p)B$
	Don't wrap	C	D	$pC + (1\text{-}p)D$
Probability		p	$1\text{-}p$	

load effect makes the model very large. A local search algorithm for finding a good but not necessarily optimal solution can be constructed by solving a smaller model iteratively. This smaller model is an efficient formulation of model 1, where the object function has adapted costs coefficients. First the smaller model is solved with starting values for the handling costs. The solution gives information on which combination of materials is used during which handling activity. Consequently the handling costs per plant can be adapted and the smaller model can be solved again. This process continues until the solution no longer changes. We suggest starting with a solution where unfavourable combinations of materials exist. These combinations contain small amounts of potted plants. Consequently the distribution costs are high. Because the model is minimized, solutions with better material combinations and lower distribution costs are found. Although optimality cannot be guaranteed, this lead to good solutions for practical problems.

14.3.5 Data

The binary programming model described in this section has been implemented in a decision support system (DSS). The use of models and data bases is controlled through a user-friendly interface. Three types of data can be distinguished. The first type concerns data selected by the user ('input data'). The second type of data consists of additional information the system needs to construct the model ('system data'). Finally the solution of the problem is translated to user-relevant information ('output data'). The following data are considered as input data. The user selects the links to build a distribution chain. At each link, the user can give the requirements on the state of the potted plant. The user also defines the size of the potted plant (height and diameter) and gives the transportation distances. Finally the user can change the default number of potted plants contained by the selected materials.

To construct a binary programming model, the system uses the following information. It needs to know at which points in the links demands have to be satisfied. The relations between demands and handling, between links and handling, between handling and material, and between material and material are retrieved from the data base. The costs of handling, material, transporta-

56*40   28*40

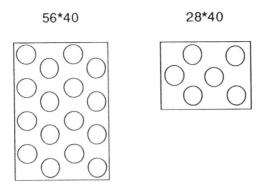

Figure 14.5   Small versus big ecotray.

tion and storage are also retrieved from the data base. Finally the system uses the numbers of potted plants in packing material, containers and trucks.

The output consists of the allocation of logistical processes to the links of the distribution chain, in the calculated sequence. The costs of the logistical processes are presented, the costs for the links are shown separately and the total distribution costs are displayed. Also the actual number of potted plants in the various material combinations can be shown.

14.4   Results

With the logistical model described, four types of problems can be tackled. First, the model can support decisions concerning the layout of the distribution chain, by allocating logistical processes in the distribution chain. Second, the model can give a better insight into tuning losses, by comparing actual distribution scenarios with distribution scenarios generated by the model. Third, the costs for alternative distribution scenarios can be calculated and compared. Fourth, the model can support decisions concerning the selection of packing and transportation materials, by calculating load effects. During the development of the DSS there were two interesting questions for VBN. First, the auctions could ask the growers to deliver a specific plant with or without a certain type of packing material. The efficiency of this requirement for the distribution chain depends on the fraction of the market which finally demands this material added to the plant and the further handling allocation in the chain. The second question was the selection and introduction of a new type of (standardized) environmentally friendly material.

14.4.1   Minimizing tuning losses under uncertainty

In Section 14.2 three types of logistical demand are described. In this example, we consider a market demand, namely the presence of plastic wrapping material. (We are dealing with a potted plant where the influence of wrapping material on product quality can be neglected.) Two types of retailers can be distinguished: retailers who think that wrapped potted plants are more attractive and consequently easier to sell, and retailers who avoid selling plastic for environmental reasons. In the distribution network under consideration there is a *decoupling point*: the flower auction. In the links before the decoupling point the destination after the decoupling point is unknown. The auction can require the presence or absence of the wrapping material by delivery on the auction. The choice between these two decision alternatives is made by minimizing the expected costs. The problem can be represented in a decision scheme (Dannenbring and Starr, 1981). This is shown in Table 14.5.

With the logistical model four optimal distribution scenarios can be generated with the corresponding distribution costs A, B, C and D. The fraction of the retailers requiring wrapping material can be determined with a market survey. This fraction is translated to the probability p that a potted plant should be wrapped at the retailer. The costs of the distribution scenarios are multiplied by these probabilities. The expected costs of the two decision alternatives are calculated and compared with each other. The decision alternative with the lowest expected costs should be implemented.

14.4.2 Selection of material on load effect

Because of the environmental legislation in some important European markets, Dutch flower auctions are developing new, environmentally friendly packing material, which we will call *ecotrays*. Besides the environmental aspects the load effect is also considered. Two formats are proposed: a small ecotray (28 cm × 40 cm) and a big ecotray (56 cm × 40 cm). With the big ecotray better load effects and consequentially lower distribution costs are achieved, because of the more favourable area–outline ratio (see Figure 14.5). However, besides a logistical unit, packing material is also a commercial unit. Retailers do not want to buy too many potted plants at a time. The maximum number of potted plants in a packing material is therefore limited. That is why small potted plants are sold in small ecotrays. For certain medium-sized potted plants it is hard to decide in which ecotray they should be packed. Commercial and logistical aspects have to be weighed. With the support of the logistical model, the logistical aspects can be quantified. As advice for VBN, we have calculated the logistical costs of an average distribution chain (Dutch grower – flower auction 'Aalsmeer' – exporter – German retailer), varying the size of the potted plant and the format of the ecotray. For a single potted plant of a certain size, we have calculated the difference in logistical costs between the use of the small ecotray and the use of the big ecotray and multiplied this by the expected amount of potted plants to be distributed in a year. The results have actually contributed to the discussion on the introduction of new, environmentally friendly and logistically pleasing packing material (Anonymous, 1993).

14.5 Conclusions

This chapter shows how the sequence of handling and choice of materials in a logistical chain can be modelled in a binary programming model in which the binary variables describe the handling and absence and presence of material on a product in the chain. The model implemented in a DSS can help to get insight into tuning losses by comparing actual distribution scenarios with optimal distribution scenarios. It can support decisions on the introduction of new material.

The DSS which has been developed for VBN to evaluate choices for the distribution network of potted plants was used for a few years until 1992. It contains data on handling costs, costs of material and relations between the materials. It was built in a data base environment in which the user can easily check and change the data. Running one optimization operation typically takes a few minutes, of which most time is consumed by data handling. Of course this amount of time depends on the number of requirements on the material on the plant in the chain and the number of links in the chain. It has been used mainly to calculate the possible consequences of introducing a new packing material as described in Section 14.4.

In this way the savings in various distribution chains on handling (packing from one material into another) have been thoroughly investigated. The advantage of the OR approach is that it gives a good overview of the necessary handling in the chain. In 1994 the ecotray as described in Section 14.4 was introduced in the potted plant logistical chain. The calculated cost savings contrast with many technical problems. The trays appear to fit badly on the containers and the plants appear to stick out too far causing physical damage.

Appendix: models 1 and 2

Model 1

$$\min \sum_{i=1}^{I} \sum_{j=1}^{J} (c_{is(j)}^{+} Y_{ij}^{+} + c_{is(j)}^{-} Y_{ij}^{-}) \tag{A14.1}$$

subject to

$X_{ip} = 0$ or $X_{ip} = 1$ for $i = 1, \ldots, I$; for all p:
decision points where demands have to be
satisfied   (A14.2)

$$X_{ij} = X_{ij-1} + Y_{ij}^+ - Y_{ij}^- \quad \text{for } i = 1, \ldots, I; j = 1, \ldots, J \tag{A14.3}$$

$$\sum_{i=1}^{I} (Y_{ij}^+ + Y_{ij}^-) \leq 1 \quad \text{for } j = 1, \ldots, J \tag{A14.4}$$

$$X_{i'j-1} + Y_{ij}^+ + Y_{ij}^- \leq 1 \quad \begin{aligned} &\text{for } i = 1, \ldots, I; i' = 1, \ldots, I; \\ &i \neq i'; j = 1, \ldots, J; h(i) \leq h(i') \end{aligned} \tag{A14.5}$$

$$X_{ij}, Y_{ij}^+, Y_{ij}^- \in \{0, 1\} \quad \text{for } i = 1, \ldots, I; j = 1, \ldots, J. \tag{A14.6}$$

Constraints (A14.2) describe the requirements made on the state of the potted plant. Constraints (A14.3) are balance equations which ensure that handling activities are performed. Constraints (A14.4) restrict the number of handling activities performed at decision point j to one, obtaining an explicit handling sequence. Constraints (A14.5) govern the application of handling activities by hierarchical relations between materials.

Model 2

$$\min \sum_{i=1}^{I} \sum_{j=1}^{J} \sum_{k=1}^{K} (c_{is(j)k}^+ Z_{ijk}^+ + c_{is(j)k}^- Z_{ijk}^-) \tag{A14.7}$$

subject to

$$X_{ip} = 0 \quad \text{or} \quad X_{ip} = 1 \quad \begin{aligned} &\text{for } i = 1, \ldots, I; \text{ for all } p: \text{ decision points} \\ &\text{where demands have to be satisfied} \end{aligned} \tag{A14.8}$$

$$X_{ij} = X_{ij-1} + Y_{ij}^+ - Y_{ij}^- \quad \text{for } i = 1, \ldots, I; j = 1, \ldots, J \tag{A14.9}$$

$$\sum_{i=1}^{I} (Y_{ij}^+ + Y_{ij}^-) \leq 1 \quad \text{for } j = 1, \ldots, J \tag{A14.10}$$

$$X_{i'j-1} + Y_{ij}^+ + Y_{ij}^- \leq 1 \quad \begin{aligned} &\text{for } i = 1, \ldots, I; i' = 1, \ldots, I; i \neq i'; \\ &j = 1, \ldots, J; h(i) \leq h(i') \end{aligned} \tag{A14.11}$$

$$Y_{ij}^+ + \sum_{i' \in \kappa_k} (X_{i'j} - 1) - Z_{ijk}^+ \leq 0 \quad \begin{aligned} &\text{for } i = 1, \ldots, I; i' = 1, \ldots, I; i \neq i'; \\ &j = 1, \ldots, J; k = 1, \ldots, K; \\ &h(i) > \max_{i' \in \kappa_k} h(i') \end{aligned} \tag{A14.12}$$

$$Y_{ij}^- + \sum_{i' \in \kappa_k} (X_{i'j} - 1) - Z_{ijk}^- \leq 0 \quad \begin{aligned} &\text{for } i = 1, \ldots, I; i' = 1, \ldots, I; i \neq i'; \\ &j = 1, \ldots, J; k = 1, \ldots, K; \\ &h(i) > \max_{i' \in \kappa_k} h(i') \end{aligned} \tag{A14.13}$$

$$X_{ij}, Y_{ij}^+, Y_{ij}^-, Z_{ijk}^+, Z_{ijk}^- \in \{0, 1\} \quad \begin{aligned} &\text{for } i = 1, \ldots, I; j = 1, \ldots, J; \\ &k = 1, \ldots, K. \end{aligned} \tag{A14.14}$$

Constraints (A14.8), (A14.9), (A14.10) and (A14.11) are equal to constraints

(A14.2), (A14.3), (A14.4) and (A14.5) in model 1. Constraints (A14.12) and (A14.13) link the variables Z_{ijk}^+ and Z_{ijk}^- to Y_{ij}^+ respectively Y_{ij}^- and X_{ij}.

References

ANONYMOUS (1993) Doorbraak fustpool en maatvoering plantenverpakkingen, *Vakblad voor de Bloemisterij*, **20**, 9.

BOWERSOX, D. J. *et al.* (1986) *Logistical management: a systems integration of physical distribution, manufacturing support and materials procurement*, 3rd Edn, New York: Macmillan.

DANNENBRING, D. G. and STARR, M. K. (1981) *Management science; an introduction*, Auckland: McGraw-Hill.

POOT, E. H. and HENDRIX, E. M. T. (1994) Optimization of potted plant distribution chains, in Beulens, A. J. M. *et al.* (eds) *Proceedings of the Second Working Conference on Optimization-based Computer Aided Modelling and Design, 1992*, pp. 143–62, Leidschendam: Lansa.

VAN DER LAAN, D. C. *et al.* (1988) *Ketenonderzoek bloemisterijprodukten, deelrapport 3: De afzetketen van bloemisterijprodukten in beeld gebracht*, Leiden: VBN.

PART FIVE

Capstone OR

Single Component	Connected Subsystem	Complex System
Operational Decisions	Tactical Plans	Strategic Scenarios
Static	Evolutionary	Dynamic
Well-Defined	Multiple Concerns	Fuzzy
Small	Large	Huge

Fingerprints

- Systems are complex and delicately balanced, like the human body or an ecological system.

- Attention is shifting from handling acute problems to designing adequate preventive measures. Hence, there is a focus on longer-term strategic scenario analysis.

- The decision environment is highly dynamic. Major changes occur frequently: compare, for example, the current map of Europe with the one in 1990.

- Goals are extremely fuzzy and the stakes can be enormous. They are not always clearly stated and can easily be influenced by hidden agendas, such as national politics or pressures from strong industrial lobbies.

- Huge databases, constantly fed by on-line monitoring systems and held by different agencies, need to be filtered, merged, aggregated and interpreted.

OR can make sense of huge quantities of data from scattered sources. This is why we termed the part 'Capstone OR'. Indeed, OR models can be superimposed upon the various data models to decide upon strategic issues such as the optimal allocation of scarce (public) resources. For instance, where to locate pollution abatement efforts in Europe given simulation results showing cause–effect relations between point pollution (e.g. smoke stacks in Russia) and regional effects (e.g. acid rain damage in Scandinavia). Various examples of OR's role in this context are given in Chapter 16, showing that OR is used by policy makers, *in casu* the European Commission, as a support in their legislative work.

Of course, acceptance of a 'mathematical' approach in fields that are traditionally averse to anything that remotely smacks of mathematics is not obvious – diplomats are more often lawyers than engineers, for example. Chapter 15 is a case in point. Preventive replacement of heart valves can be very beneficial, the OR model says. Will this lead the national health service, or any other insurer, to subsidize such interventions in order to reduce costs in the long run? OR expertise is not yet accepted in the world of politics and in court, compared, for example, with engineering expertise. Nevertheless, there is a wealth of new applications here, widening the scope of our discipline as well as its relevance to society.

CHAPTER FIFTEEN

Reoperations on patients with possibly defective artificial heart valves

The Björk–Shiley Dilemma

J. H. P. VAN DER MEULEN, E. W. STEYERBERG, L. A. VAN HERWERDEN, Y. VAN DER GRAAF and J. D. F. HABBEMA

15.1 The environment

During the last 50 years medical practice has been through a metamorphosis. The scientific knowledge that has been built up in the medical field during that period is vast. There has been an explosion of knowledge and the number of scientific articles that are published each year in the medical field grows exponentially. As a consequence the diagnostic and therapeutic arsenal has expanded very quickly too. Next to the growth in knowledge we can identify other developments that have strongly influenced medical practice. First, societal developments leading to individualization and emancipation are linked to a more critical and demanding attitude of the public towards the medical professionals and their activities. Second, efforts to contain increasing health care costs make medical professionals and policy makers more inclined to consider the outcome of medical procedures in relation to their costs. It is in this rapidly changing field that decisions have to be made about the appropriateness of diagnostic and therapeutic procedures on various levels. Some decisions have a societal reach, like vaccination programmes for infants or screening programmes for breast and cervical cancer. Others address institutional problems (in hospitals and other institutions) and involve for instance the prevention of infections and prophylactic treatment to prevent deep venous thrombosis in immobilized patients. Finally, there are decisions that relate to the treatment of individual patients or defined groups of patients with a specific disease.

Not very long ago the general attitude in medicine was that each patient is unique and that the doctor knows best. A thorough knowledge of physiology and pathology together with personal observations from clinical experience were thought to provide sufficient guidance for medical decision making. During the last two decades, however, this attitude has rapidly given way to a new approach, which is lately referred to as 'evidence-based medicine'. In this approach decisions are as much as possible based on scientific evidence about the effectiveness of medical procedures. In other words, decisions are preferably based on clinical experience with patients that is recorded explicitly in a scientific (i.e. reproducible and unbiased) way. A development that follows naturally from this line is the introduction of methods to support rational decision making in medicine. The first article that explicitly described the application in medical diagnosis of methods that originate from operational research and decision theory appeared in 1959 (Ledley and Lusted, 1959). From that time on the introduction of decision theory in medicine went on gradually. In 1980 the first textbook appeared (Weinstein *et al.*, 1980) followed by a journal entirely devoted to the methods and applications of decision theory in medicine (*Medical Decision Making*). At present, decision analyses and cost effectiveness studies appear frequently in most leading medical journals. The number of clinicians who are really familiar with these methods and techniques is rather small. Most published applications have been performed by epidemiologists, mathematicians or economists. In the majority of cases, however, the initiative to perform the study was taken by clinicians and the collaboration between clinicians and decision scientists is an essential element in a successful completion.

The major benefit of the introduction of decision theory in medicine is not and will not be an *ad hoc* application in the consulting-room or at the bedside. It is rather exceptional that a decision analysis will be performed directly linked to everyday patient care. Methods and techniques from decision theory have become very important, however, in clinical research and education. Decision analyses provide an excellent format to give a synthesis of the available scientific evidence with respect to the effects of alternative diagnostic or therapeutic actions on the future health of patients. They may lead to the formulation of explicit clinical practice guidelines. Teaching decision theory to medical students and clinicians will make them aware of the general principles that underlie diagnostic and therapeutic decision making.

In this chapter we will give an example of how a close collaboration between clinicians, epidemiologists and experts in clinical decision sciences led to explicit recommendations with respect to the decision of whether to reoperate or not on patients with a possibly defective mechanical heart valve prosthesis. Parts of the following sections were published in a different form in *Circulation*, a major journal in cardiology (Van der Meulen *et al.*, 1993).

15.2 The problem

In the early 1960s prosthetic heart valves were introduced as a treatment option for patients who suffered from a stenosis or insufficiency of the cardiac valves between the left atrium and ventricle (mitral valve) or between the left ventricle and the aorta (aortic valve). Over the years manufacturers have changed and revised the specifications of these valves to get optimal haemodynamic characteristics, long durability and a low risk of complications like thromboembolism (obstruction of a blood vessel by a blood clot that originates from the prosthetic valve). In the mid 1980s the Björk–Shiley convexo-concave (BScc) heart valve prosthesis was withdrawn from the market after repeated reports of outlet strut fracture. At that time about 86 000 patients had received this valve worldwide: about 82 000 with an opening angle of 60° and about 4000 with an opening angle of 70°. By November 1991, 466 outlet strut fractures had been reported to the manufacturer, which has to be considered as an underestimate of the true incidence (Pfizer, 1992a).

In 1991 the results of a retrospective cohort study were reported that provided detailed information on all patients in The Netherlands with a BScc valve (Van der Graaf et al., 1992). It describes the experience of 2588 BScc valves implanted in 2303 patients between 1979 and 1985, followed up for a mean of 6.6 years. Information on vital status was obtained from municipality registers and information about the cause and mode of death was obtained from the patient's general practitioner or retrieved from clinical records. The yearly risk of strut fracture appeared to be constant over time. It was demonstrated that the risk was greater for larger valves (≥ 29 mm), for valves with an opening angle of 70°, for valves implanted in the mitral position and for valves of younger patients.

Shortly after the results of this study were made public the staff of the Department of Cardiopulmonary Surgery of the University Hospital Rotterdam asked the Centre for Clinical Decision Sciences of the Erasmus University in Rotterdam to support them in identifying those patients who would benefit from prophylactic replacement of this possibly defective heart valve prosthesis. At that time prophylactic replacement was recommended only for patients with large (≥ 29 mm) 70° BScc valves of an early production series (group I 70° BScc valves) (Lindblom et al., 1989). This multidisciplinary group evaluated the effects of prophylactic replacement using prognostic information obtained from the Dutch follow-up study. For each valve type, the age of the patients was determined, below which replacement is beneficial. Also the cost-effectiveness of replacement was studied as a function of the patient's age.

15.3 The OR approach

The analytical approach required structuring of the clinical problem for decision analysis and estimating the essential probabilities. The structuring

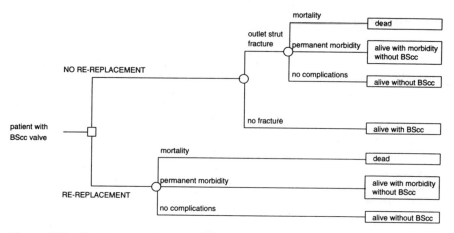

Figure 15.1 Decision tree for prophylactic replacement of the Björk–Shiley convexo-concave (BScc) heart valves. A representation of the four possible health states and the transitions among them is given on the right.

of the decision problem was rather easy, because the decision to reoperate on patients with the possibly defective heart valve prosthesis corresponds with an archetypical problem in medicine. The treatment decision requires that a short-term (surgical) risk together with all its accompanying discomfort and anxiety for the patients is balanced against a lifelong ongoing risk (of mechanical failure of the heart valve prosthesis) with its potentially disastrous consequences. The same structure can be recognized for instance with respect to the routine decision to perform coronary artery bypass surgery in patients with an atherosclerotic stenoses in the vessels that supply blood to the cardiac tissue. The required probabilities were derived from the retrospective Dutch cohort study using multivariate regression techniques.

15.3.1 Structure of the decision model

The structure of the problem is represented by the decision tree in Figure 15.1. The model contains four health states ('alive with a BScc valve', 'alive without a BScc valve', 'alive with severe morbidity without a BScc valve' and 'death'). A discrete-time Markov process was applied to represent the patient's prognosis as a sequence of particular states of health and possible transitions among them during time intervals of one year (Beck and Pauker, 1983). These survival estimates were used to calculate the expected lifetime a patient will spend in each of the health states. The calculations were performed with computer software especially designed for medical decision problems (Decision Maker, New England Medical Center, 1988). For mathematical details see appendix.

15.3.2 Probability estimates

The probability estimates required for this analysis were (1) the surgical mortality and morbidity after prophylactic replacement, (2) the age-specific annual risk of death, (3) the annual risk of outlet strut fracture, and (4) the mortality and morbidity after strut fracture. If possible, these probabilities were derived from the retrospective cohort study including 2303 Dutch patients with a mean duration of follow-up of 6.6 years (Van der Meulen *et al.*, 1993). We reanalysed the data from this study using logistic regression and Poisson regression to derive prognostic models for the surgical mortality, age-specific risk of death and the risk of outlet strut fracture. Variable selection was performed with a forward stepwise procedure based on the significance level of the partial likelihood ratio test (limit for significance to enter 0.10).

15.3.2.1 *Surgical mortality and morbidity of prophylactic replacement*

Data about the risks associated with prophylactic replacement of artificial valves is scarce (Wideman *et al.*, 1982; Husebye *et al.*, 1983). We defined surgical mortality as death occurring during the first 30 days after surgery and we assumed it to be equivalent with the 30 day mortality after primary valve replacement. The surgical mortality after replacement was then estimated with a logistic regression model. Surgical mortality for a 40-year-old patient without any risk factor is 1.5 per cent. Age (odds ratio, 1.02 for each additional year), a BScc valve in the mitral position (odds ratio, 2.6), concomitant bypass surgery during valve replacement (odds ratio, 1.5), acute endocarditis (odds ratio, 2.2), a poor ventricular function (odds ratio, 2.9) and valve replacement as emergency treatment (odds ratio, 6.3) are incremental risk factors. Information on the risk of permanent severe morbidity after valve surgery was derived from a follow-up study on neurological complications of coronary bypass surgery and estimated to be 1.3 per cent (Shaw *et al.*, 1986).

15.3.2.2 *Age-specific annual risk of death*

In general, the life-expectancy of patients with mechanical heart valves is lower than that of the general population (Blackstone and Kirklin, 1985; Kirklin and Barratt-Boyes, 1986). To obtain age-specific mortality rates we carried out Poisson regression for death after primary valve replacement. Patients with mechanical failure of the valve prosthesis were considered censored observations. The annual risk of death for a female patient younger than 40 without any risk factor is 0.6 per cent. Age (hazard ratios for patients between 40 and 49, 1.3; between 50 and 59, 2.2; between 60 and 69, 3.6; between 70 and 79, 7.3), concomitant bypass surgery (hazard ratio, 1.5), a

BScc valve in the mitral position (hazard ratio, 1.6) and male gender (hazard ratio, 1.3) are incremental risk factors. From this Poisson model we approximated the age-specific annual risk of death for patients after valve replacement. This approximation is based on the assumption that mortality after the operative period depends on the attained age and the condition of the patient rather than on the time elapsed since valve replacement. The age-specific hazard rates were assumed to be constant for patients younger than 35, while those for patients older than 80 were estimated on the basis of exponential extrapolation.

15.3.2.3 *Annual risk of outlet strut fracture*

The results of the Dutch follow-up study indicate that the annual risk of strut fracture is constant over time and depends on valve characteristics and age at implantation. We performed Poisson regression to estimate these effects on the annual risk of strut fracture. The annual risk of fracture is 0.09 per cent in patients younger than 40 with a small 60° aortic valve. Age at implantation (hazard ratios for patients between 40 and 49, 0.4; over 50, 0.3), a valve prosthesis in the mitral position (hazard ratio, 3.3), a large valve size (hazard ratio for valve size ≥ 29 mm, 3.8) and an opening angle of 70° (hazard ratio, 5.8) are incremental risk factors.

15.3.2.4 *Mortality and morbidity after outlet strut fracture*

A patient sustaining a mechanical failure of the valve may die immediately or after an attempted emergency valve replacement. The mortality after aortic strut fracture is high: in the Dutch cohort study, six out of seven reported patients died (86 per cent). Mortality after mitral strut fracture was lower: of the 35 reported patients 18 died (51 per cent). These mortality rates were adopted in the present analysis. It was assumed that 50 per cent of the survivors of an outlet strut fracture will have severe permanent morbidity. This estimate is based on an evaluation of the functional status of the Dutch patients who survived the outlet strut fracture.

15.3.3 Outcomes

We calculated life-expectancy with and without replacement of the BScc valve. To account for the fact that most patients are risk averse we investigated the effects of discounting future life-years at 5 per cent per year (McNeil *et al.*, 1978) and also the effects of adjusting for the quality of life by weighing the time spent with severe permanent morbidity from valve surgery or outlet strut fracture with a quality adjustment factor of 0.5 (each year for a patient with severe morbidity is worth half a year in full health).

The direct medical costs were represented for replacement and expectant management in 1990 Dutch guilders (Dfl) (Vrieze *et al.*, 1992; Financial Statistics, 1990; Van Hout, 1990) (1 US dollar is approximately 2 Dfl). The costs of prophylactic replacement were estimated to be Dfl 20 000 (angiography, surgery, three days' intensive care and 10 days' low care). The costs of an outlet strut fracture amount to Dfl 15 000 for a patient who dies after admission to hospital (surgery and five days' intensive care). Furthermore, it was assumed that 50 per cent of the patients who die after an outlet strut fracture die outside the hospital. The costs for patients who survive after an outlet strut fracture are Dfl 45 000 (surgery for valve replacement, five days' intensive care and 20 days' low care followed by surgery for removal of the fractured strut, one day intensive care and 12 days' low care). Future costs were also discounted to the present value at a 5 per cent per year discount rate.

15.3.4 Simplifications

Our analysis was subject to the following simplifying assumptions: (1) the surgical mortality of prophylactic replacement is equivalent to the 30 day mortality after primary valve replacement; (2) survival of patients with an artificial valve is determined by their attained age and clinical condition and not by the time elapsed since primary valve replacement; (3) replacement of the BScc valves obviates the risk of strut fracture without affecting long-term mortality and morbidity; (4) the annual risk of strut fracture is constant over time; (5) the possibility that any mechanical heart valve (whether of the BScc type or not) may have a finite life span, even without strut fracture, is not considered in our analysis.

15.4 Results

To illustrate the results of the decision analysis we start by presenting the effects of prophylactic replacement for a fictitious 40-year-old male patient with a large 60° mitral BScc valve without comorbidity. According to the prognostic models presented earlier we estimated for this patient the annual risk of strut fracture to be 1.1 per cent, surgical mortality to be 3.7 per cent and the 'basal life-expectancy', i.e. the life-expectancy if the strut fracture risk is assumed to be zero, 25.0 years. The life-expectancy of this patient increases from 23.1 to 24.1 years if prophylactic replacement is performed. So, prophylactic replacement adds 1.0 years (or 4.3 per cent) to the life-expectancy. Expressed in terms of loss, prophylactic replacement gives rise to a 53 per cent reduction of the loss in life-expectancy that is attributable to outlet strut fracture (1.0 from 1.9 years).

The main results of this study originated from sensitivity analyses. First,

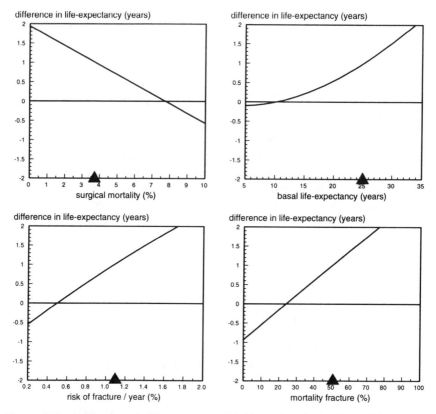

Figure 15.2 Difference in life-expectancy with and without prophylactic replacement as a function of the quantitative estimates for a 40-year-old male patient with a 29 mm 60° mitral Björk–Shiley convexo-concave heart valve. A positive difference indicates that the life-expectancy with prophylactic replacement is higher than without. The triangles indicate the estimates we used for this particular patient.

the quantitative estimates were varied one by one over wide ranges (see Figures 15.2 and 15.3). It appeared that all probability estimates have a substantial independent effect on the difference in life-expectancy with and without prophylactic replacement. Variations in the preferences for the length (i.e. discounting future life-years) and quality of life have a less distinct effect. In Figure 15.4 the age thresholds for the prophylactic replacement of BScc valves are shown for male and female patients without comorbidity. For each valve type the age thresholds were calculated, first using simple future life-years, second using discounted life-years and third using discounted and quality-adjusted life-years. Confidence intervals were calculated for these age thresholds using the upper and lower limits of the 95 per cent CI of the estimated strut fracture risk. The effect of discounting future years of life and

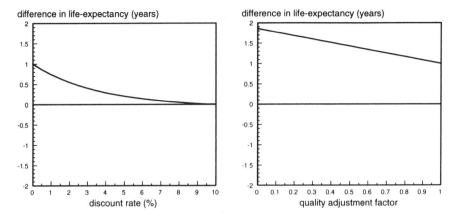

Figure 15.3 Difference in life-expectancy with and without prophylactic replacement as a function of the yearly discount rate of future life-years and of the quality adjustment factor for permanent morbidity caused by prophylactic replacement or strut fracture. (Same patient as in Figure 15.2.)

adjusting for the quality of life on the age thresholds is relatively small for most valve types. Only for small 60° valves either in the mitral or aortic position did the age thresholds decrease considerably when discounted future life-years are used. In Figure 15.5 the age thresholds for male patients with and without comorbidity are presented and if 1 per cent or 3 per cent is added to the surgical mortality as well as to the risk of permanent morbidity after prophylactic replacement.

In Figure 15.6 we display the marginal cost-effectiveness of prophylactic replacement for male patients without comorbidity as a function of age for the various valve types. The costs per discounted and quality-adjusted life-year gained depend upon valve type and position. Replacement of mitral BScc valves produces lower cost-effectiveness ratios than replacement of aortic valves, indicating that replacement is more cost effective. The costs per discounted and quality-adjusted life-year gained rise steeply as the patient's age approaches the threshold for replacement. Repeat analyses for females gave similar results.

15.5 Implementation

The results of the decision analysis indicate also that patients with other BScc valve types than the formerly identified large (≥ 29 mm) 70° BScc valves of an early production series (group I 70° BScc valves) may benefit from prophylactic replacement. Prophylactic replacement may also increase the (discounted and quality-adjusted) life-expectancy in patients with other valve types as is represented in Figures 15.4 and 15.5. This required recommenda-

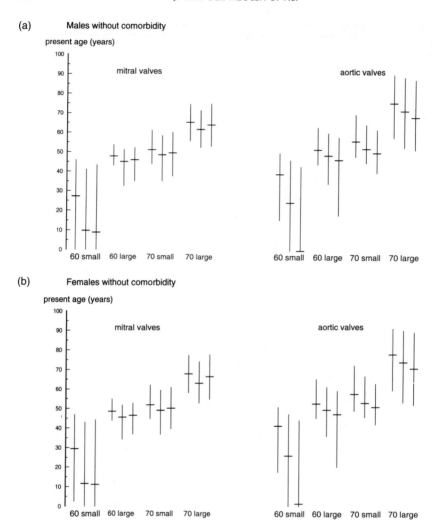

Figure 15.4 Age thresholds for prophylactic replacement in (a) male and (b) female patients. For each valve type age thresholds are based on simple life-years, discounted life-years and discounted and quality-adjusted life-years (from left to right). The thresholds indicate the age of the patient below which prophylactic replacement increases the life-expectancy. The vertical bars indicate confidence intervals (CI) which were calculated using the upper and lower limits of the 95 per cent CI of the strut fracture rates.

tions that depend on the clinical profile of the patient. The main interest of the application of decision theory in medicine is that decisions can be augmented by having explicit and quantitative estimates of the net benefit of the alternative options available. Furthermore, sensitivity analyses indicate the most influential factors of the decision problem.

(a)   Males without comorbidity

present age (years)

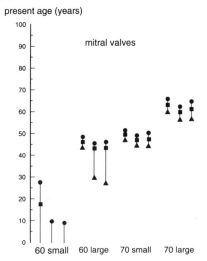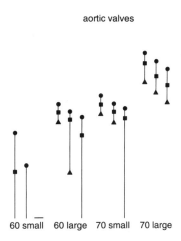

(b)   Males with poor ventricular function

present age (years)

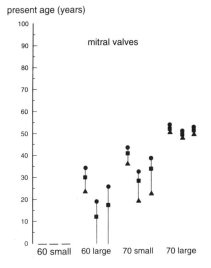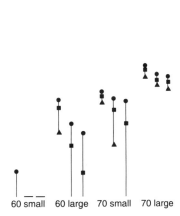

Figure 15.5   Age thresholds for prophylactic replacement in male patients (a) without comorbidity and (b) with a poor left ventricular function. These figures present the age thresholds, if 0 (circle), 1 (square), or 3 per cent (triangle) is added to the surgical mortality and morbidity. The surgical mortality is estimated with a logistic model as a function of age and clinical risk factors (see Section 15.3) and to this risk estimate an extra risk of 1 or 3 per cent is added. For each valve type age thresholds are based on simple life-years, discounted life-years and discounted and quality-adjusted life-years (from left to right). A horizontal line at age 0 indicates that no age threshold could be found.

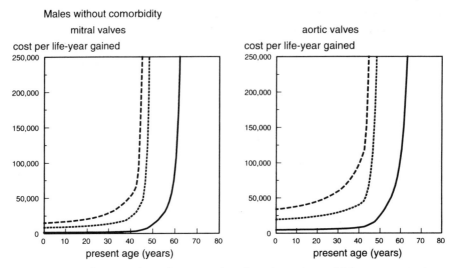

Figure 15.6 The costs per discounted and quality-adjusted life-year gained as a function of age for male patients with large 60° (dashed line), small 70° (dotted line) and large 70° (solid line) mitral and aortic valves. Costs are expressed in 1990 Dutch guilders (1 US dollar was approximately 2 Dutch guilders).

The results of a decision analysis can never dictate the clinical management of individual patients. There is inevitably a 'gap' between the results of a formal decision model and the actual clinical situation with its immeasurable amount of specific clinical features. The definitive decisions always remain the responsibility of the clinician. The implementation of the results of a decision analysis must always follow this basic principle. First, oral presentations were given for interested cardiologists and cardiac surgeons. The aim of these sessions was to present the main results of the analysis, especially the age thresholds as presented in Figure 15.4, and at the same time to clarify the principles of clinical decision making for this particular Björk–Shiley dilemma. During these sessions it was intended to provide clinicians with a conceptual framework that could be used for decision making for all their patients with these possibly defective heart valves in their own practice. Second, we estimated on clinicians' request the (discounted and quality-adjusted) life-expectancy for individual patients. Cardiologists and cardiac surgeons were asked to provide us with the essential clinical information (i.e. sex, age at implantation, valve type and clinical profile), which allowed an individualized estimation of the probabilities from the logistic and Poisson regression equations. The results of the decision analysis were presented to the clinicians together with a written explanation and sometimes with results of sensitivity analyses as presented in Figure 15.2. Third, we condensed the results of the study into a scientific publication with emphasis on results rather than on methodology and understandable for interested clinicians without any specific knowledge of decision theory (Van der Meulen *et al.*, 1993). At the end of 1993 22 patients were reoperated on in The Netherlands (De Mol *et al.*, 1994). Unfortunately we do not know exactly to what extent this decision analysis contributed to the decision to reoperate.

15.6 State of affairs

This study illustrates how decision analysis can be used to summarize the results of a detailed epidemiological study in a form that is relevant for clinical decision making. In this particular example, multivariate regression models could be used to estimate the required probabilities for specific clinical profiles. This is seldom possible, however; more often quantitative estimates have to be derived from the medical literature and sometimes even from the personal experience of experts. It is a remarkable experience that the explicit and quantitative results of this decision model for individual patients were practically always consistent with the clinical judgement of cardiologists and cardiac surgeons. This is an important indication of the validity of the decision analytic approach.

The Björk–Shiley dilemma is a worldwide problem (Birkmeyer *et al.*, 1992; Blackstone and Kirklin, 1992). Not surprisingly other groups also published the results of a decision analysis with respect to this problem. An informal analysis of these publications reveals that the results of all these studies are more or less similar and that they disagree only on details. On the one hand this could be expected, because the studies used partly the same epidemiological data. On the other hand, however, it supports the validity of the application of decision theory for this clinical problem, because the studies were performed completely independent of each other.

One specific element of the problem is that the risk of mechanical failure of the valve may depend on possible manufacturing defects, which are related to specific valve types and to the manufacturing (weld) date (De Mol *et al.*, 1994; Pfizer, 1992b; Woodyard, 1991). During the course of time more and more information about these possible manufacturing defects will become available, which will necessitate a continuous updating of the replacement decision for individual patients. It is for this reason that a publication is prepared that provides clinicians with a 'paper and pencil' tool (in the form of a small number of easy-to-use tables and diagrams), which enables them to estimate quantitatively the effect of replacement on the prognosis of an individual patient. At this moment it is not possible to comment on the impact of such a tool, but clinicians and patients may benefit by having available a tool that can be applied whenever new information about the risk of mechanical failure of the Björk–Shiley convexo-concave mechanical valve prosthesis becomes known in the future.

Acknowledgements

Figures and part of the text of the article 'Age thresholds for prophylactic replacement of Björk–Shiley convexo-concave heart valves', originally published in *Circulation*, 1993, Vol. 88, pp. 156–64, are published with permission of the American Heart Association.

Appendix: mathematical details

The probabilities of being alive with a BScc valve ($P_{BS,i}$), of being alive without a BScc valve ($P_{OV,i}$), and of being dead at the end of the ith cycle ($P_{dead,i}$) are respectively given by

$$P_{BS,i} = P_{BS,i-1}(1 - m_i)(1 - f_i)$$

$$P_{OV,i} = P_{OV,i-1}(1 - m_i) + f_i(1 - \tfrac{1}{2}m_i)(1 - I)P_{BS,i-1}$$

$$P_{dead,i} = P_{dead,i-1} + m_i(1 - \tfrac{1}{2}f_i)P_{BS,i-1} + f_i(1 - \tfrac{1}{2}m_i)IP_{BS,i-1} + m_iP_{OV,i}$$

where m_i is the probability of death during the ith cycle, f_i is the probability of outlet strut fracture, and I is the probability of death after fracture. The expected duration of life in health state j, LE_j, is the weighted sum of durations of all possible cycles, the weight being the probabilities of being in that state during the cycles:

$$LE_j = \frac{1}{2}\sum_{i=1}^{n}(P_{j,i-1} + P_{j,i})t_i,$$

where t_i is the duration of the ith cycle (one year). It is assumed that the transitions occur in the middle of each cycle. In the final model the health state 'alive with severe morbidity without a BScc valve' has been added.

References

BECK, J. R. and PAUKER, S. G. (1983) The Markov process in medical prognosis. *Medical Decision Making*, **3**, 419–58.

BIRKMEYER, J. D., MARRIN, C. A. S. and O'CONNOR, G. T. (1992) Should patients with Björk-Shiley valves undergo prophylactic replacement? *Lancet*, **340**, 520–3.

BLACKSTONE, E. H. and KIRKLIN, J. W. (1985) Death and other time-related events after valve replacement, *Circulation*, **72**, 753–67.

(1992) Recommendations for prophylactic removal of heart valve prosthesis, *Journal of Heart Valve Disease*, **1**, 3–14.

DE MOL, B. A., KALLEWAARD, M., VAN HERWERDEN, L. A., DEFAUW, J. J. and VAN DER GRAAF, Y. (1994) Single-leg strut fractures in explanted Björk-Shiley valves, *Lancet*, **343**, 9–12.

FINANCIAL STATISTICS (1990) (in Dutch), Utrecht: Nationaal Ziekenhuisinstituut.

HUSEBYE, D. G. *et al.* (1983) Reoperation on prosthetic heart valves, *Journal of Thoracic and Cardiovascular Surgery*, **86**, 543–52.

KIRKLIN, J. W. and BARRATT-BOYES, B. G. (1986) *Cardiac Surgery*, New York: John Wiley.

LEDLEY, R. S. and LUSTED, L. B. (1959) Reasoning foundations of medical diagnosis, *Science*, **130**, 9–21.

LINDBLOM, D., RODRIGUEZ, L., BJÖRK, V. O. (1989) Mechanical failure of the Björk-Shiley valve, *Journal of Thoracic and Cardiovascular Surgery*, **97**, 95–7.

MCNEIL, B. J., WEICHSELBAUM, R. and PAUKER, S. G. (1978) Fallacy of the five-year survival in lung cancer, *New England Journal of Medicine*, **229**, 1397–401.

PFIZER/Shiley Heart Valve Research Centre (1992a) Dear Doctor letter, March. (1992b) Dear Doctor letter, September.

SHAW, P. J., BATES, D., CARTLIDGE, N. E. F., HEAVISIDE, D., FRENCH, J. M., JULIAND, D. G. and SHAW, D. A. (1986) Neurological complications of coronary artery bypass graft surgery: six month follow-up study, *British Medical Journal*, **293**, 165–7.

VAN DER GRAAF, Y., DE WAARD, F., VAN HERWERDEN, L. A. and DEFAUW, J. (1992) Risk of strut fracture of Björk-Shiley valves, *Lancet*, **339**, 257–61.

VAN HOUT, B. A. (1990) Heart transplantation; costs, effects and prognosis, Thesis (in Dutch), Erasmus Universiteit Rotterdam.

VAN DER MEULEN, J. H. P., STEYERBERG, E. W., VAN DER GRAAF, Y., VAN HERWERDEN, L. A., VERBAAN, C. J., DEFAUW, J. J. A. M. T. and HABBEMA, J. D. F. (1993) Age thresholds for prophylactic replacement of Björk-Shiley convexo-concave heart valves. A clinical and economic evaluation. *Circulation*, **88**, 156–64.

VRIEZE, O. J., BOAS, G. M. and JANSSEN, J. H. A. (1992) *An econometric model for a scenario analysis of coronary heart disease* (in Dutch), Maastricht: Rijksuniversiteit Limburg.

WEINSTEIN, M. C. and FINEBERG, H. V. (1980) *Clinical Decision Analysis*, Philadelphia: W. B. Saunders Co.

WIDEMAN, F. E., BLACKSTONE, E. H., KIRKLIN, J. W., KARP, R. B., KOUCOUKOS, N. T. and PACIFICO, A. D. (1982) The hospital mortality of re-replacement of the aortic valve. Incremental risk factors, *Journal of Thoracic and Cardiovascular Surgery*, **82**, 692–8.

WOODYARD, C. (1991) Firm told to warn 350 with heart valves, *Los Angeles Times*, 28 April.

CHAPTER SIXTEEN

OR and the environment

A Fruitful Combination

P. VAN BEEK, L. FORTUIN, L. N. VAN WASSENHOVE
and L. HORDIJK

16.1 Introduction

Operational Research (OR) is a discipline which primarily aims at improving the effectiveness and the efficiency of the processes of decision making. These processes occur in every segment of our society: industry, banking, agriculture, government, politics, etc. The distinctive feature about OR is that it commonly makes use of optimization models. From the very start of the 1980s these models have been increasingly 'cloaked in user-friendly wrappings', the so-called decision support systems. In previous publications the authors have already indicated the potential significance of OR for society (Fortuin *et al.*, 1989a,b). Using a number of daily recurrent environmental issues, the following will illustrate how OR can be used in describing and solving them.

Things are not going well for the environment. This general and maybe blunt-sounding statement is pre-eminently valid for a country that likes to characterize itself as 'The Netherlands, Land of Distribution', but which at the same time is being urged by society as well as politics to keep a sharp eye on the protection of the environment. The tension between profiting from economic activity, reducing the deficit on the national government's budget and expansion of the public sector, on the one hand, and concern for a good social climate, for both our children and our grandchildren, on the other, is getting stronger every day. There is a tendency to give economic growth the highest priority. Therefore it is becoming more and more important to increase the effectiveness of environmental measures.

OR can play an important role in visualizing and solving environmental problems. This observation is at the same time an implicit appeal to increase the contribution of OR to the variety of environmental research carried out. Furthermore the subject of OR should get more attention in the different environmental courses at Dutch universities and students of OR should be encouraged to apply their experts' appraisal to environmental problems.

16.2 Environmental problems on a world scale

On Thursday 19 October 1989 the Economics Supplement of *NRC Handelsblad* (a high-quality Dutch newspaper) contained the following articles: 'Bonn finances environmental research out of privatization', 'Notorious environmental polluters out of Iran' and 'Building of manure factory delayed'. The frequent publication of environment-related contributions of this kind is bound to grow. In that very same week the magazine *SAFE* contained the article 'Assistance from space to help the environment'. In this article pictures taken by NASA astronauts distressingly show that the ecological balance of our Earth is in danger, owing to human activity (overpopulation, environmental pollution, excessive use of energy). NASA has devoted more than a hundred million dollars to get the project 'Mission to Planet Earth' off the ground. By means of observations during satellite flights the dynamics of the Earth can be sufficiently fathomed. Consequently environmental disasters can be predicted and there will be time to find ways to avert them.

16.2.1 Acidification in Europe

At the International Institute for Applied System Analysis (IIASA) in Austria a study of the *acidification issue* in Europe (Hordijk, 1988; Shaw, 1988) has been carried out. IIASA developed the RAINS model (Regional Acidification INformation and Simulation) as a tool for analysing the consequences of various energy scenarios for Europe. The various submodels of RAINS are based on mathematical–physical relations describing air, soil and water movements, and on chemical relations describing the transformation of species. Using RAINS, people can project the deposition of SO_2, NO_x and NH_3 within Europe when the location and size of the emission sources are known (Figure 16.1). The model translates the calculated deposition values into effects on the environment.

RAINS is completely interactive: the user selects an energy scenario for his or her point of departure, decides upon which emissions he or she would like to calculate, for which future years, which sets of measures will have to be implemented per country and per year and which output will have to be generated. The output of RAINS largely consists of maps of Europe showing for example deposition patterns and the excess of deposition over specified environmental targets (so-called critical loads; Hettelingh *et al.*, 1996).

Next, OR becomes involved, as the question is how, with limited financial resources, one can reach a previously agreed environmental objective, with

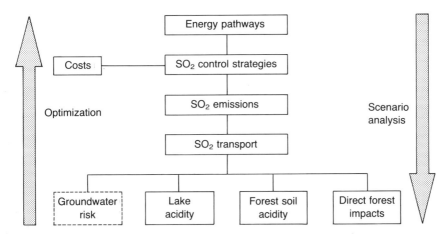

Figure 16.1   Diagram of the RAINS model. (From Alcamo *et al.*, 1987.)

the best results. This question can also be put differently: try, within available budgets, to invest in emissions reduction in such a way as to realize a maximum effect on the environment. For sulphur (S) the environmental objective can then be translated into grams per square metre per year. In order to make optimizations of this kind, RAINS has been extended with an optimization module.

RAINS can be used for several aims. First of all, it offers the possibility to show on a map of Europe what the SO_2 deposition pattern will look like in the year 2020, when no SO_2 emission reductions have taken place. The places with maximum sulphur deposition would then be mainly located in Central and Eastern Europe. The highest peaks are the Donetz Basin in the southeast of the European part of the Ukraine, the area around Leipzig and Dresden in the former GDR, and the Krakow/Katowice territory in the south of Poland. The peak load here is more than 10 grams S per m^2 per year. In Western Europe the high deposition areas would then be situated in northern Italy, the Ruhr area in Germany, and in central England. Here the values are lower than in Eastern Europe, but they are still considerably above the maximum value permitted for ecosystems in Europe. The same exercise can be done for future years, taking into account agreements concerning emissions reduction on a European level. After the 1985 international agreement to reduce acid rain some 12 billion guilders have been set aside for this reduction. Notwithstanding these actions, we still notice a fairly high peak load of 7.5–10 grams S per m^2 per year.

In order to optimize truly, this amount of 12 billion guilders per year has been taken as a starting-point. This amount has to be spent in such a way as to minimize the peak load. Analysis with the help of RAINS demonstrates that with that sum a peak load of 4.5 grams S per m^2 per year will be possible. However, in order to reach that minimum, the money will have to be allocated in a different way, namely 8.2 billion guilders in Eastern Europe

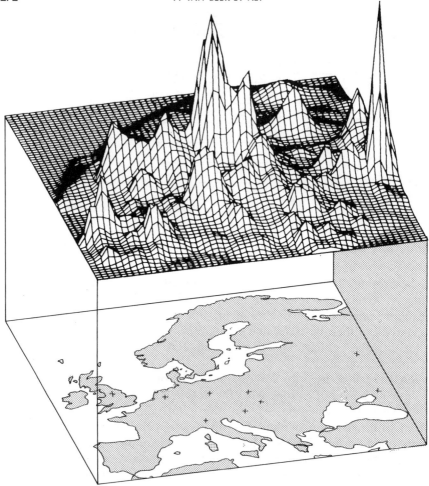

Figure 16.2 A three-dimensional picture of sulphur deposition in Europe in 1980, produced by the RAINS model. Crosses indicate the 10 locations of the highest deposition points. (From Hordijk, 1988.)

and 3.8 billion guilders in Western Europe. In other words, assuming the previous allocation of the resources for the purpose of reducing SO_2 emission (divided on a 50–50 basis), a shift of 2.2 billion guilders from Western to Eastern Europe will lead to a halving of the peak load! As an illustration, Figures 16.2 and 16.3 show the sulphur deposition in Europe for the years 1980 and 2000.

The optimization module in RAINS can also be applied in a different manner. Suppose the peak load has to be reduced to 3.0 grams S per m^2 per year. How much money will then be needed every year, assuming an optimal allocation of money? A similar question was posed by the European negotiators who discussed a new sulphur emission reduction protocol. After

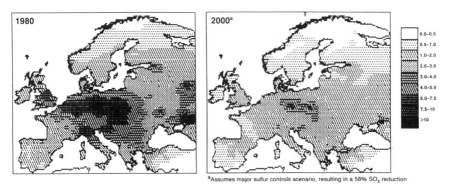

Figure 16.3 Total sulphur deposition in Europe, in 1980 and 2000 (grams per m^2 per year). (From Alcamo *et al.*, 1987.)

some years of discussion the negotiators agreed on a common goal for environmental protection: the deposition of sulphur should be less than or equal to 40 per cent of the difference between the 1990 deposition in each point of Europe (RAINS uses a 150 × 150 km grid) and the critical loads in those points. This so-called 'gap closure scenario' was accepted as the best compromise for the short term, knowing that reaching deposition equal to the critical loads would be very costly, and in some places could not be reached at all. The RAINS model was used to calculate the cheapest possible way to reach these targets. In the course of the negotiations a large set of cost minimization runs were carried out: 50 per cent gap closure, twice the value of the critical loads as target values, maximum emission reduction percentages for some East European countries, etc. As shown in Table 16.1 the new sulphur protocol (signed in Oslo, June 1994) is largely based on the results from the RAINS model. The total cost of the emission reductions agreed in the protocol for the year 2000 ranges between 20–25 billion DM when add-on technology is applied, using the prices in 1990 and the energy structure of 2000 as a basis. The risk of environmental damage in 2010 is reduced to 10% of the European area showing sulphur deposition values exceeding the critical loads for sulphur.

After the success of the European RAINS model, the World Bank commissioned Wageningen Agricultural University, IIASA and five other partners to build a similar model for Asia. This model has not yet been used in negotiations, because the political structure needed for such negotiations is lacking in Asia. However, many countries are now using the RAINS-ASIA model to define their position as an acid rain 'receptor or donor' and comparing these figures with neighbouring countries.

Concluding, it can be stated that decision support systems such as RAINS can be used to:

- evaluate various scenarios, thereby visualizing effects such as 'peak load' of the SO$_2$ deposition geographically;

Table 16.1 SO$_2$ emissions for European countries as calculated with RAINS and agreed in negotiations based on RAINS

Signatories	2000	2005	2010	RAINS	Close^a
Austria	80	80	80	80	*
Belarus	38	46	50	38	
Belgium	70	72	74	77	*
Bulgaria	33	40	45	74	
Croatia	11	17	22	40	
Czech Rep.	50	60	72	72	*
Denmark	80	80	80	72	*
Finland	80	80	80	80	*
France	74	77	78	80	*
Germany	83	87	87	90	*
Greece	0	3	4	0	*
Hungary	45	50	60	68	
Ireland	30	30	30	41	
Italy	65	73	73	73	*
Liechtenstein	75	75	75	—	
Luxembourg	58	58	58	58	*
Netherlands	77	77	77	77	*
Norway	76	76	76	76	*
Poland	37	47	66	66	*
Portugal	0	3	3	0	*
Russian Fed.	38	40	40	38	*
Slovakia	60	65	72	72	*
Slovenia	45	60	70	45	*
Spain	35	35	35	55	
Sweden	80	80	80	83	*
Switzerland	52	52	52	52	*
UK	50	70	80	79	*
Ukraine	40	40	40	56	

^{a}*indicates that emissions reduction in 2010 is larger than RAINS results or at maximum three percentage points lower.
Sources: UN Protocol ECE/EB.AIR/40 (1979); Amann and Schöpp (1993).

- show the effects of investments in the reduction of acidifying emissions;
- optimize investment allocations, thereby aiming to reach prespecified environmental targets.

16.2.2 The greenhouse effect

At the moment the Dutch National Institute for Public Health and Environmental Protection (RIVM) is developing a simulation model which

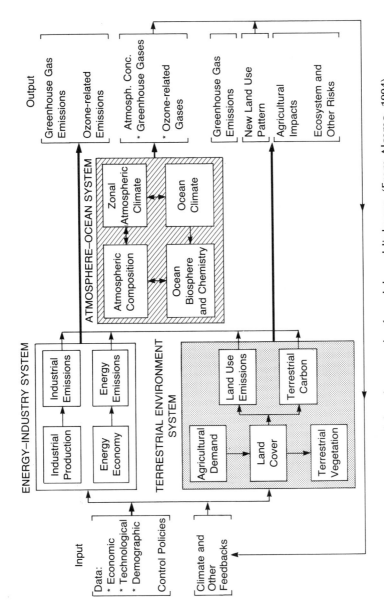

Figure 16.4 The IMAGE model: a framework of models and linkages. (From Alcamo, 1994).

is to give a clearer understanding of the mechanisms behind *the greenhouse effect* (Rotmans *et al.*, 1989; Alcamo, 1994). This model, known as IMAGE (Integrated Model for the Assessment of the Greenhouse Effect), aims to give an integrated view of the greenhouse effect, as well as providing an insight into the basic mechanisms of this phenomenon. It is based on a large quantity of data, obtained from the literature and by consulting many experts. Experimental research is also playing a role here. All information is being integrated and brought to the same level of aggregation. The central feature is the interweaving of knowledge from various sciences.

The core of IMAGE consists of an energy/industry system (for calculating emissions), a terrestrial environment system (providing land-use emissions) and the atmosphere–ocean system (calculating atmospheric concentrations). These modules are linked in such a way that the outcome of one module serves as the input for the next. In the greenhouse effect it is especially the gases CO_2, CH_4, N_2O, CFC-11 and CFC-12 that play a part (Figure 16.4). The IMAGE 2.0 model covers the whole world (subdivided into 13 geographical regions) at a 0.5° latitude by 0.5° longitude level.

Historical emissions during the years 1900–85 have been stored on computer; for the period 1990–2100 four emission scenarios have been selected: *Conventional Wisdom*, *Biofuel Crops*, *No Biofuels* and finally the *Ocean Realignment* scenario. The different scenarios are based on a study of both the anthropogenic sources of trace gas emission (such as energy usage, agriculture and industry) and of the growth of world population and natural sources.

The first scenario (*Conventional Wisdom*) assumes a continuation of current economic growth with no climate-related policies. The *Biofuel Crops* scenario assumes that a substantial amount of biofuels comes from new cropland. In this scenario the assumptions about the development of the economy and energy growth are the same as in the first scenario. The third scenario assumes that *No Biofuels* are used and are replaced by oil. Finally, the *Ocean Realignment* scenario investigates the consequences of a major change in the ocean's circulation pattern. This scenario is an example of a surprise scenario.

IMAGE shows that the Montreal Protocol (an agreement concerning the restriction of CFC gases (UNEP, 1987)) is important in order to stabilize the relative contribution of CFCs to the greenhouse effect.

From the system of international agreements, such as the Montreal Protocol, standards for the emission of CFCs are laid down. Models like IMAGE can predict the effects of this standardization for a longer period and can therefore play an important part in determining (international) standards. OR will, to an important extent, be able to help in finding effective measures to approach the appointed standards as close as possible and with the lowest financial effort. In the near future, a model for optimizations of this kind will be added to IMAGE. However, it has to be stated that application of the formulated optimization criterion may be difficult, since quantifying the costs related to long-term environmental effects is often hardly possible.

16.3 Three Dutch cases

In the light of three examples of Dutch applications, we will now have a closer look at the role of OR.

16.3.1 The manure issue

Owing to legal measures against top dressing, people will, in future and much more than now, have to deal with the transport, storage and processing of manure on a large scale. Since the costs involved are bound to increase, it is important to select the means of transport, the depots and the processing installations in such a way that the costs will be kept at a minimum. In order to gain some more insight into this logistic issue work is being done on a decision support system as part of a cooperative project between IMAG (Institute for Mechanization, Labour and Construction) and the Wageningen University of Agriculture (Faculty of Mathematics). This system is based on a mathematical model which determines an optimal logistic structure in a real-life situation, namely the production of manure and the possibilities for disposal, transport, storage, etc. By changing the starting points and subsequently determining once again the optimum structure, the effect of the different starting points becomes clearer.

Because of the growth of intensive stock breeding it has become much harder to find an acceptable destination for the amount of manure that is being produced. This has led to the top dressing of estates, and subsequently groundwater pollution. The introduction of legal measures has highlighted this problem and it is bound to present even more difficulties in future.

The core of the problem is the surplus of minerals (phosphates, nitrates and potassium) and heavy metals (copper, cadmium and zinc) in the animal manure. The balance between the supply and drainage of minerals in agriculture has been disrupted by the use of feeders from outside the farm. Importing feed produce for intensive stock breeding plays a major part here. As a result of the surplus of minerals it will no longer be possible to dispose of the amount of animal manure within the agricultural sphere, at least, not in a sensible way. An overloading of animal manure will affect soil fertility and lead to deteriorating quality of the crops. Furthermore, this will affect the quality of the groundwater and contribute to the eutrophication of the surface water in areas poor in nourishment. Ammoniac emission from manure is one of the causes of acid rain.

Possible ways of disposing of animal manure are restricted because of the *phosphate standards.* In practice often more manure will be produced than is legally permitted. The surplus has to be taken off the market – fully or partially. This is what we call 'processing'. In doing so, other manure products can be produced. For example, after the purification of calf manure, one is left with calf manure silt which is being disposed of in agriculture. Processing pig manure on an industrial scale, as two Dutch firms, Promest in Helmond and MeMon in Deventer, are planning to do, yields an organic granular manure substance comparable with artificial fertilizer which should be brought onto the market at a similar price.

While discussing *disposal* and *processing*, aspects like (means of) transport, storage and manufacturing are important. In order to enlarge the disposal prospects for animal manure, a number of treatments are possible, such as separation, stumming, sedimentation and drying. The destination for manure that has been treated is disposal. Most of the treatment techniques are also applied in processing. The use of means of transport, storage facilities, processing installations, etc., is necessary in order to give the amount of animal manure a destination – disposal or processing. There are many logistic possibilities in achieving this. It is not clear *a priori* which ones are to be preferred.

Since 1987, IMAG and the Wageningen University of Agriculture (Faculty of Mathematics, OR Section) have been working on a computer system which could provide an insight into this logistic issue. This decision support system will, in specific cases, determine an optimal logistic structure, including transport and storage needs. A structure like this is dependent on the given possibilities. By gradually altering these and by again calculating the optimal structure, the effects of the adaptations will become visible. The system treats the issue on a regional level; there are two areas that are worked on: a surplus area and a shortage area. An example of a possible combination of a surplus and a shortage area is the combination of the Dutch provinces North Brabant and Zealand.

As has been mentioned, the destination of manure can be either disposal or processing. In determining the optimal ways of disposal and processing, three steps can be distinguished: (1) disposal on one's own farm, (2) disposal on other farms in the same region, (3) disposal in a different region as well as processing. Most of the manure can probably be disposed of on one's own farm. The disposable quantity is calculated by considering the production of manure and the disposal possibilities for each farm. The utilization of existing possibilities is also estimated. Since every farmer is autonomous there are no further optimizations involved in this step. Instead it tries to calculate the real situation as well as possible.

After disposal on every farmer's own farm has been determined, for each municipality the amount of manure that can be disposed of on other farms within the same region is determined. This step, too, is based on an estimation of the actual use of manure. If there is still a surplus after these steps have been taken, the question arises of whether this can be disposed of in an area which has a shortage. If this is not the case, the only possibility that remains is processing.

For this situation the flows of manure will be optimized. For each kind of manure available destinations have to be indicated and in what way they can be realized in terms of storage, transport, etc. It goes without saying that the costs also have to be indicated.

The structure of the flows of manure strongly depends on the possibilities. In the case of direct transport from surplus farms in a surplus area to shortage farms in a shortage area, the flows will have a simple structure. But the structure will be more complicated in the case of central storage in either a surplus or a shortage area and in the case when transport involves transfer (due to the use of different means of transport). The structure of the flows of manure is then modified in a *network* with flows of manure going through branches to different nodes in the network. A branch represents the transport of manure products, a node is either a point of departure and/or a destination for a branch and can visualize a place of storage or a possible processing facility (Figure 16.5).

The flows in the network are optimized within the existing possibilities, in which case use is made of *linear programming* (LP). The following conditions are taken into account in solving this LP problem:

- all surpluses have to have a destination (disposal or processing);
- the flows going to a processing installation (or place of storage) must not exceed its capacity;
- the inflow at a place of storage (or place of transfer) is equal to the outflow;
- if processing involves the production of manure products, the size of this production will be related to the size of the flow to the processing installation;
- total disposal must not surpass the available storage space for manure;
- disposal of a certain kind of manure is limited, in the same way as the disposal to other farms in the same area.

All conditions are linear. If the locations of storage facilities, for example, are not predetermined, this option can be included in the optimization, just like scale effects regarding storage and processing.

At the moment the decision support system for the manure issue, BOSMEST (Beslissing Ondersteunend Systeem voor MESTproblemen), is complete. Part of it is used at the National Manure Bank. The optimal logistic structure can be determined in any real situation. Furthermore BOSMEST can be used to determine the correct standards and the cost of the accompanying logistic structure. Therefore politicians can also use it in order to agree on useful standards.

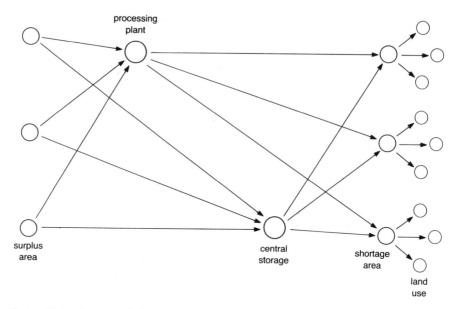

Figure 16.5 A network for storage, transport and processing of manure.

16.3.2 Soil clearance projects

The clearance of Dutch soil is a socially sensitive subject which, from the beginning of the 1980s, has involved a lot of money and other resources. In dealing with the clearance of Dutch soil the provinces play a central part under the Interim Law for Soil Clearance. All provinces are in charge of monitoring cases of soil pollution, establishing priorities for research and clearance, supervising and/or carrying out research and clearance together with dealing with their financial completion.

The amount of money that the provinces can spend on soil clearance (ranging from a few to several tens of millions of guilders each year) is far too small to achieve a near conclusion of the operation. The soil clearance programme, drawn up yearly by the provinces, indicates which cases of soil pollution are qualified for research or clearance in the coming year. It also indicates the urgency of the projects. The ways in which the degree of urgency has been determined varies for each province; they are executed in accordance with the Interim Law for Soil Clearance.

The urgency-determining systems differ in three ways: in the extent of detail, in the criteria that are taken into account, and in the extent to which the system is of a quantitative or qualitative kind. The soil clearance policy pursued by the provinces aims for the selection of projects and clearance variants to be carried out in such a way that the budget available for soil clearance will be exploited in the best possible way.

For each case of soil pollution several possibilities exist for handling the project; these possibilities are translated into clearance variants. The variants consist of applying a soil clearance technique or a combination of techniques. In describing these techniques one can distinguish between those without preceding clearance and those that are applied after clearance has taken place. One can also distinguish between purifying and non-purifying techniques.

Users of the system have gained a lot of experience with the techniques that imply clearance and transport to a cleaning installation. The techniques to be applied without clearance have been tried less often; some of them are still in the development phase. Because of the lower cost and the often less radical activities involved in the latter group of soil clearance techniques, these are bound to be applied more often in future. However, the problem here is that the degree of soil clarification that will eventually be reached is often hard to check, or checks are insufficient because of inhomogeneous clarification.

The environmental effect of carrying out a project according to a specific clearance variant is determined by comparing two variables: (1) the risk of the pollution for the national health, and (2) the same risk after clearance has taken place. As a matter of course the use of the soil and the kind of clearance techniques to be applied are important factors.

The risk of pollution is determined on the basis of criteria which are related to the nature and the concentration of the pollutants, to the extent of the pollution, to the possible ways of distribution and to the risks of exposure. It is the nature of the location in particular which is normative for the use of the soil. In selecting the clearance techniques, implementation, effectiveness and reliability play the key roles. These three aspects are determined by way of a scoring system; the user assesses the score of the criteria that have been included by way of assigning weight to the different criteria.

The aim of soil clearance can be described as selecting combinations of 'project/clearance variant' in such a way as to maximize the sum of the environmental effects (total environmental effect), without overrunning the budget.

This objective has been translated into an LP model with integer variables. The possibility exists of including an additional objective: namely, that a specific number of projects, minimal or maximal, will be carried out. The addition of objectives of this kind to the first objective has been translated into the so-called goal programming model. Determining the environmental effects and cost per variable and maximizing the total environmental effect together constitute the allocation model. The model has been applied to eight projects and corresponding clearance variants in the province of Overijssel. The LP model can be transcribed into mathematical terms. The problem is solved with the help of the computer packages LINDO and SCICONIC.

Conclusively it can be stated that:

■ The allocation model can be a useful aid in pursuing a soil clearance policy on a provincial level. On a national level the system could be used to determine the projects to be carried out nationally, and consequently to determine the distribution of the budget among the provinces. The advantage is that on this level the progress of soil clearance will be less influenced by non-predictable factors than on a provincial level.

■ The system as a whole can be computerized so that the effects of different policy views and the chosen limiting conditions can be calculated in an efficient and simple way.

16.3.3 Optimization models in water treatment technology

In the last 20 years people in The Netherlands have worked hard to reduce the effects of waste water drainage on the quality of surface water. As a result hundreds of sewage treatment plants have been installed. Only in a few cases have engineers been able to make an optimal choice of purification techniques and designs of these RWZIs (water purification installations), for the optimal choice depends on the specific limiting conditions that characterize each situation. These are enforced, among others, by:

■ the amount and quality of the waste water to be purified;
■ the accessibility of (new) technologies;
■ the requirements for the quality of the effluent;
■ the location of the RWZI;
■ the advantages and disadvantages of the technologies that are being applied.

As an example we mention the model which was used during the 1970s in a certain province for the planning of new sewage treatment plants. In close collaboration with an engineering firm, the provincial department of public works had made a study to find the cheapest way of purification, from a provincial point of view. The cheapest alternative appeared to be purification by means of a so-called oxidation ditch. Hence it was decided to dig a number of oxidation ditches all over the province. However, a few years later it appeared that the extension of an oxidation ditch for the removal of phosphates by means of a tertiary water purification stage would cause problems. The only suitable method for the removal of phosphates from the effluent of an oxidation ditch is simultaneous dephosphatization by way of precipitation through a chemical reaction. The amount of sludge this generates is considerable, while at the same time the processing possibilities of this sludge are small because of the very high percentages of heavy metals.

In view of the efforts the government will make in the coming years to accomplish a reduction of the phosphate percentage in the effluent to be drained, and in view of the problems described here, it seems justified to conclude in this case that the decision makers have not chosen the *optimal* way of purification. In retrospect it becomes clear that they should have selected a method of purification which takes into account the specific disadvantages of an oxidation ditch, i.e. hardly any possibilities for extension, neither in capacity nor in application of tertiary purification.

It is likely that in the future there will hardly be any new sewage treatment plants installed in The Netherlands. However, the existing RWZIs will have to be expanded or renovated because of age, low capacity or negative secondary environmental effects. Examples of such effects are eutrophication due to phosphate draining, a high production of sludge, stench, etc. At the moment most sewage treatment plants are not fit to remove the phosphates from the surface water. In the years to come a tertiary purification stage will therefore have to be installed in many cases. For the time being the following variants are suitable for the removal of phosphates: chemical precipitation, a granular reactor, a granular reactor combined with biological dephosphatization, and magnetic separation. The Dutch administrators of water quality had to decide upon the way in which they would comply with more rigid phosphate standards, before these standards came into force in 1992.

The financial consequences of the strict measures taken by the national government can best be illustrated with an example: the Hoogheemraadschap Rijnland (Top Dike Board Rhineland) will have to budget 70 million guilders. The other administrative bodies will have to pay comparable amounts.

Like the other administrators of water quality, Rijnland will have to answer the following questions:

- Which method for dephosphatization, available at the moment, is optimal?
- Which locations are optimal for the tertiary purification stages?
- Is it a matter of a few bigger or many smaller dephosphatization installations?

The qualification 'optimal' in the first two questions is dependent on the balance between the following criteria:

- cost (investment cost and yearly cost);
- primary environmental effects (results of the various methods for the removal of phosphates);
- secondary environmental effects (odour inconvenience for people living in the neighbourhood, production of sludge) and the potential measures the government may consequently take (e.g. as part of the Nuisance Act);

- amount of space;
- stability and reliability;
- possibilities of extension (flexibility related to future policy);
- knowledge of different techniques, required from the contractor, but particularly from the operator.

16.3.4 Expert systems

In determining the optimal method of dephosphatization and the optimal location(s), simulation models and optimization models can be used. With these models engineers have tried to describe a real situation and with some models they have even succeeded in obtaining a better insight into that reality.

In the last few years information technologists have made much progress in the development of so-called 'expert systems', also called 'knowledge systems'. Compared with a human expert, the major advantage of such a system is the large amount of knowledge it can handle. Moreover, the system, being computer based, cannot overlook or forget any knowledge, as long as that knowledge has been entered in the correct way. Furthermore it offers the possibility of integrated use (at several places) of quantifiable knowledge (data) and non-quantifiable knowledge. Improving and extending existing sewage treatment plants is an extremely complicated matter with great financial consequences. In view of the complexity and because of the large amount of knowledge this requires, an expert system is an obvious instrument to solve the problem. At present researchers at the Wageningen Agricultural University are initiating the development of expert systems for water treatment technology.

In summary it can be stated that OR can play an important role in the development of water purification systems. Optimization models show the best location for treatment installations and the technology to be used. Furthermore OR models, when integrated in expert systems, are playing a role which is becoming more and more important.

16.4 Conclusion

The chapter has shown, with the help of examples, how and where OR can be applied in environmental science. It goes without saying that the examples were chosen arbitrarily, and are far from exhaustive.

We conclude this discussion with a list of environmental problem areas where OR has been or can be applied, namely:

- drafting environmental security systems;
- drafting reservoir systems and their control;

- water quality control;
- location policy for waste removal installations;
- determining environmental risks;
- location policy for polluting industrial installations;
- optimization of emissions from industrial installations.

These examples are derived from Marchuk (1986) and Pinter (1987). Both publications contain a wealth of other examples.

Acknowledgements

This chapter could not have been written without the contribution and the cooperation of our colleagues: namely, R. J. Swart (RIVM) who gave us information on the IMAGE model (Rotmans et al., 1989); R. M. de Mol (IMAG) who (together with the Agricultural University) developed the BOSMEST system (de Mol, 1989), A. J. Jacobse and P. P. G. Wolbert who discussed the subject of 'Budget allocation in soil clearance' in their Master's theses at the Wageningen Agricultural University (they received the Unilever Research Price 1988 for their work) (Jacobse and Wolbert, 1988); and E. C. Doekemeijer who supplied us with information concerning the design of the simulation and optimization models for water treatment installations (Doekemeijer, 1989).

References

ALCAMO J. et al. (1987) Acidification in Europe: a simulation model for evaluating control strategies, Ambio, 16, 232–45.

ALCAMO, J. M. (ed.) (1994) IMAGE 2.0 Integrated Modeling of Global Climate Change, Dordrecht: Kluwer.

AMANN, M. and SCHÖPP, W. (1993) Reducing excess sulphur deposition in Europe by 60 percent, Background Paper for the UN/ECE Working Group on Strategies, International Institute for Applied Systems Analysis, Laxenburg.

DOEKEMEIJER, E. C. (1989) Rapport Haalbaarheidstudie optimalisatiemodel Zuiveringstechnologie; Rapport Vakgroep Waterzuivering, Landbouwuniversiteit Wageningen.

FORTUIN, L., VAN BEEK, P. and VAN WASSENHOVE, L. N. (1989a) Operationele Research kan meer voor u doen dan u denkt, De Ingenieur, 101(3), 22–8.

(1989b) Logistiek: meer dan een modewoord, Harvard Belgium Review (20), Third Trimester, 54–7.

HETTELINGH, J-P., POSCH, M., DE SMET, P.A.M. and DOWNING, R. J. (1996) The use of critical loads in emission reduction agreements in Europe, Water, Air and Soil Pollution (in press).

HORDIJK, L. (1988) A model approach to acid rain, Environment, 30(2), 17–42.

JACOBSE, A. J. and WOLBERT, P. P. G. (1988) Saneringsproject en saneringstechniek: een keuzeprobleem, Afstudeerverslag Vakgroepen Cultuurtechniek en Wiskunde, Wageningen, februari.

MARCHUK, G. I. (1986) *Mathematical Models in Environmental Problems*, Amsterdam: North-Holland.

DE MOL, R. M. (1989) A decision support system to optimize the application and processing of manure, *Land and Water Use, Agricultural Engineering*, **1**, 409–15.

N.N. (1989) Hulp vanuit de ruimte moet milieu helpen, *SAFE*, oktober, 62–7.

PINTER, J. (1987) A conceptual optimization framework for regional acidification control, *Systems Analysis and Model Simulation*, **4**, 213–26.

Protocol to the 1979 Convention on Long-range Transboundary Air Pollution on Further Reduction of Sulphur Emissions, ECE/EB.AIR/40, New York: United Nations.

ROTMANS, J., DE BOOIS, H. and SWART, R. J. (1989) IMAGE: an integrated model for the assessment of the greenhouse effect, Report 758471009 RIVM.

SHAW, R. W. (1988) Transboundary acidification in Europe and the benefits of international cooperation, Proceedings of a conference on Pollution knows no frontiers: priorities for Pan European Cooperation, Varna, 16–20 October.

UNITED NATIONS ENVIRONMENT PROGRAM (1987) Montreal protocol on substances that deplete the ozone layer: Final Act, Montreal.

PART SIX

The process of OR

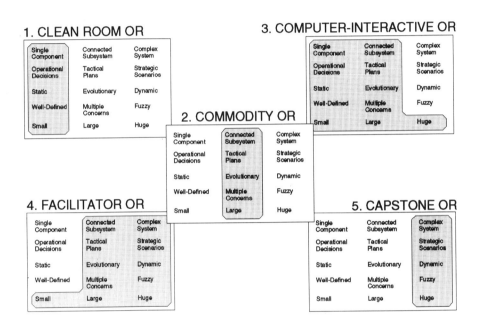

1. CLEAN ROOM OR

Single Component	Connected Subsystem	Complex System
Operational Decisions	Tactical Plans	Strategic Scenarios
Static	Evolutionary	Dynamic
Well-Defined	Multiple Concerns	Fuzzy
Small	Large	Huge

2. COMMODITY OR

Single Component	Connected Subsystem	Complex System
Operational Decisions	Tactical Plans	Strategic Scenarios
Static	Evolutionary	Dynamic
Well-Defined	Multiple Concerns	Fuzzy
Small	Large	Huge

3. COMPUTER-INTERACTIVE OR

Single Component	Connected Subsystem	Complex System
Operational Decisions	Tactical Plans	Strategic Scenarios
Static	Evolutionary	Dynamic
Well-Defined	Multiple Concerns	Fuzzy
Small	Large	Huge

4. FACILITATOR OR

Single Component	Connected Subsystem	Complex System
Operational Decisions	Tactical Plans	Strategic Scenarios
Static	Evolutionary	Dynamic
Well-Defined	Multiple Concerns	Fuzzy
Small	Large	Huge

5. CAPSTONE OR

Single Component	Connected Subsystem	Complex System
Operational Decisions	Tactical Plans	Strategic Scenarios
Static	Evolutionary	Dynamic
Well-Defined	Multiple Concerns	Fuzzy
Small	Large	Huge

Fingerprints

- In clean-room OR, process skills are confined to the ability to interact with experts from other disciplines, e.g. chip designers.

- Commodity OR requires attention for input and output links to other systems as well as foolproof design of the 'commodities'. Indeed, much like recently available statistical software, OR tools can easily be misused by inexperienced or insufficiently trained people.

- Computer-interactive OR requires a careful balancing act between attention for the needs of the users (knowledge engineering), clever design of computer interfaces, and smooth integration of (custom-made) algorithms and experience.

- In facilitator OR the practitioner acts as an integrator and facilitator between individuals or groups from different fields and with different backgrounds. Process skills are at the core of the OR intervention.

- Capstone OR requires practitioners to cooperate successfully with experts from very different fields (econometricians, engineers, economists, social scientists, etc.) as well as to be able to design and sell their approach as a strong support in complex, strategic policy-making processes.

Hence, process skills are always present but their importance and character vary with the application at hand. No matter how hard a case writer tries to incorporate process issues into the case description, it will never be a true representation of the complexities of an OR intervention. Therefore, Chapter 18 in this part attempts to capture some of these complexities based on interviews with successful practitioners (several of whom contributed to earlier chapters of this book) and their clients.

The other chapter in this part revisits the so-called 'crisis' debate. Many issues in that debate relate to the process of OR and to the differences between the application parts, e.g. tool versus process orientation. Together, these two chapters shed additional light on the challenges and pitfalls of successful OR practice and on the choices to be made with respect to the future directions in which our discipline should evolve. This, of course, has consequences for the role to be played by professional societies as well as for the design of adequate OR curricula.

CHAPTER SEVENTEEN

Crisis? What crisis?

Four Decades of Debate on Operational Research*

C. J. CORBETT and L. N. VAN WASSENHOVE

17.1 Introduction

Over the years, the operational research/management science (OR/MS) literature has shown a growing interest in the history of the field, but also a growing concern about its future. Much has been written about the future of OR/MS, claiming that future to be bright, expressing some worries or simply stating that the future is past and that OR/MS is dead. Given that so much is written about what is called the 'current crisis in OR/MS', it is reasonable to ask to what extent this debate is truly justified.

Surprisingly, writings on the OR/MS crisis generally show little or no awareness of opinions expressed in the management literature. In this chapter, we attempt to step outside this largely inward-looking debate by frequently referring to articles from the *Harvard Business Review (HBR)*, the management journal most read by executives. Although *HBR* obviously cannot be said to represent management attitudes in general, tracing its OR/MS-related articles provides some interesting insights. To begin with, as Figure 17.1 shows, the crisis debate in the OR/MS literature took off soon after a dramatic drop in attention paid by *HBR* to OR/MS. This suggests that managers are hardly interested in OR/MS any more and/or that the OR/MS community is no longer paying attention to the managerial literature, both of which would be cause for concern. However, a glance through recent editions of the practice-oriented OR/MS journal *Interfaces* suffices to demonstrate that this decline in attention in general management journals

---

*Reprinted with permission from *Operations Research*, Vol. 41, Iss. No. 4, 1993, © 1993, Institute for Operations Research and Management Sciences. No further reproduction permitted without the consent of the copyright owner.

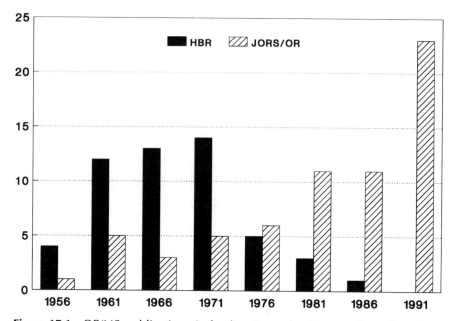

Figure 17.1   OR/MS publications in leading journals. The number of articles on OR/MS in the *Harvard Business Review* contrasted with the number of articles expressing concern about the development of the profession in the *Journal of the Operational Research Society* and in *Operations Research*, the journals of the British and American OR societies respectively. Obviously, this graph should not be taken too literally, but the overall picture remains suggestive. Remark that it was in 1973, and more viciously in 1979, that Ackoff pronounced OR/MS dead.

does not coincide with a lack of relevant and highly successful applications of OR/MS. It is surprising that the apparent drop in managerial attention has occurred despite the vastly enhanced implementation possibilities offered by the advent of computers, decision support systems, etc.

The aim of this chapter is to identify the main issues in the OR/MS debate, also taking into consideration the views expressed in *HBR*. By doing so we hope to give an impression of what are perceived as problems in OR and of how the OR community has been concerned with these problems during the past four decades. The papers referred to are generally written by leading practitioners and academics in OR, and are usually based on their personal opinions or practical experience. A different perspective is provided in Chapter 18, where many of the same issues are discussed again but on the basis of interviews with a number of practitioners and their clients. Obviously neither chapter contains the final and total truth; they should, rather, be seen as complementing each other. For a more detailed discussion of the crisis, the debate, and the underlying causes, we refer to a companion paper which has appeared elsewhere (Corbett and Van Wassenhove, 1993).

17.2 The literature

This section contains a systematic discussion of a selection of writings in the OR/MS debate, comprising both early and recent articles in the OR/MS literature and articles from *HBR*. In order to distinguish opinions from the OR/MS literature from those found in *HBR*, most of the references in the text include a mention of the journal involved. We focus on five main issues of the debate:

- tool orientation versus problem orientation in OR/MS;
- client relations in OR/MS;
- the interdisciplinary nature of OR/MS;
- the relevance of OR/MS at a strategic level;
- the learning effect of an OR/MS study.

Section 17.2.1 considers the first of these issues, which is fundamental and lies at the heart of much of what has been written; it has many implications for the issue of relations with clients (Section 17.2.2), which are complicated by the interdisciplinary nature of much OR, discussed in Section 17.2.3. Whether or not OR is and should be involved in strategic decision making (Section 17.2.4) is a sensitive point, closely related to the problems in client relations. Finally, particularly in strategic cases, OR may well have more to offer as a way of increasing one's insight into complex situations rather than as a way of finding 'the' answer to a well-defined problem; this is further elaborated in Section 17.2.5.

17.2.1 Tool orientation versus problem orientation in OR/MS

OR was born during World War II, when teams of scientists drawn from a wide range of disciplines (biology, chemistry, physics, mathematics, statistics, psychology, and others) were asked to advise on questions such as the allocation of the limited number of RAF fighter planes available over the British Isles and the Continent. At that time, OR was entirely oriented towards real-world problems; there simply were no 'OR tools'. In the early days, there was a rush to develop the much-needed tools. That this development has been extremely successful cannot be disputed, as demonstrated by the wealth of techniques OR workers currently have at their disposal. However, many of the contributors to the OR/MS debate seem to think that the focus has shifted too far toward 'Analysis in Wonderland', and away from the real world. Below, we first illustrate that excessive tool orientation is perceived as a real problem in OR; in the second section we add our own perspective on the issue.

17.2.1.1   The rule of the tool, or managing messes?

A fundamental issue in the 'crisis' debate in the OR/MS literature is that of the orientation of the field; one's individual views on most issues inevitably boil down to whether one sees OR/MS primarily as a knowledge-oriented science or as a problem-oriented technology.

The problem of the split identity of OR/MS was noted early by Herrmann and Magee (1953):

> As an applied science, the work is torn between two objectives: as applied it strives for practical and useful work; as science it seeks increasing understanding of the basic operation, even when the usefulness of this information is not immediately clear.

Drucker (1959) recognized the need for both a knowledge orientation ('*The first need of a management science is, then, that it respect itself sufficiently as a distinct and genuine discipline*', p. 30; author's italics) and also a problem orientation ('*The second requirement for a management science is, then, that it takes its subject matter seriously*', p. 148; author's italics).

Some writers very early worried that the balance in OR had already swung too far away from applications towards pure science. King (1967) expressed his concerns:

> it seems apparent that the long-range acceptance and development of the field intrinsically depends on the solution of problems in the real world and that in turn, such successes can be achieved in quantity only if we can develop and maintain a cadre of practitioners who are competent in mathematics *and* who display the creativity and ingenuity that is so important to the 'real world' aspects of problem solving.

Beged-Dov (1966) agreed:

> Although they seem to pay lip-service to applications, it is quite easy to establish that the majority of the operations researchers best known in the field are openly proud of their learned papers and almost apologetic when referring to applications.

Wagner (1971) also warned that there may be too much emphasis on techniques in OR, and too little on the problems they are supposed to solve. Ackoff (1973) argued that the mathematical tools used in OR/MS are out of date, and no longer capable of dealing with the messes faced by most managers. In his words: 'accounts are given about how messes were murdered by reducing them to problems, how problems were murdered by reducing them to models, and how models were murdered by excessive exposure to the elements of mathematics'. Morse (1977, p. 188), the first

President of the Operations Research Society of America, signals a trend away from the real world:

> But if we stick to our original aim of matching our models to reality, rather than trying to make reality fit the preferred model, we can contribute in important ways to many of the serious problems facing this country and the world.

Bishop (1979, p. 154) also observed excessive tool orientation: 'The modus operandi of the operations researcher is to abstract from some real-world problem a mathematical problem for which he can find an answer. He then touts that answer as a guidepost to the manager.'

Ackoff's (1979a) 'The future of operational research is past' is a well-known and outspoken article in the OR/MS debate. He holds academic OR and the relevant professional societies primarily responsible for what he perceives as the decline of the discipline; they brought about that (p. 94) 'OR came to be identified with the use of mathematical models and algorithms rather than the ability to formulate management problems, solve them, and implement and maintain their solutions in turbulent environments'. In other words (p. 95), 'In the first two decades of OR, its nature was dictated by the nature of the problematic situations it faced. Now the nature of the situations it faces is dictated by the techniques it has at its command.' He adds (p. 95): 'OR has been equated by managers to mathematical masturbation and to the absence of any substantive knowledge or understanding of organizations, institutions or their management'. Managers have lost interest in OR, because (p. 100) '[T]he unit in OR is a problem, not a mess. Managers do not solve problems; they manage messes.' As suggested by the titles of his papers (a more recent one was named 'OR, a post mortem'), Ackoff's views are not uncontroversial, but the ongoing debate shows that many others share his concerns.

17.2.1.2   Is tool orientation really so bad, and why?

Many of the critics claim that OR/MS tends to abstract too much from real-world problems, in order to allow more analytical sophistry. The general (though not universal) warning against letting tools become the dominant orientation in OR/MS (the rule of the tool) seems justified, as this requires potential clients to be familiar with the OR/MS toolbox in order to decide whether or not to commission an OR study. This way, OR/MS teams will only receive projects for which they already have the tools, thereby limiting growth opportunities for OR/MS in both theoretical and practical directions. Daniel (1987) points out that 'the adaptive OR group itself has proved to be the most enduring and useful "tool" of all'.

Of course, the importance of tools should not be underestimated either: the profession would never have got very far without them, nor will it get much further without continuous development of new tools. Weingartner

(1987) signals that researchers will no longer consider individual problems as being independent of each other, but attempt to develop solutions that can be used more than once. This development of general tools is necessary; designers, for example, also use general tools to create a unique design. Several examples of this are discussed elsewhere in this book; think, for instance, of the asset–liability management system described by Boender (Chapter 13), or the packages for cutting problems discussed by Schepens (Chapter 5).

In some cases clients get in touch with an OR consultant because they need the consultant's OR skills; in other cases it is because the consultant has the necessary expertise and tools for the problem situation faced by the client. The literature generally suggests that OR should be problem oriented rather than tool oriented. We believe that the issue is not so much one of choosing either extreme, but rather of finding a balance between the two. Obviously, no OR consultant interested in keeping a job can afford to lose sight of the client's problem, but at the same time, few clients are prepared to accept the high cost and risk involved in developing a solution from scratch; a tool which has already proven its value elsewhere is often a much more attractive prospect. Once an OR consultant has successfully developed and implemented a solution for one client's problem, it will often make more sense (and money) to seek out other clients with similar problems rather than to start from scratch on a completely different problem (Corbett et al., 1995). This, however, presupposes entrepreneurial tendencies and marketing skills not always found in OR workers.

An important, but not often recognized, aspect of OR/MS tools, is noted by Blumstein (1987): 'We must recognize, however, that our patents on the tools we like to think of as our own have largely expired.' It is true that tools such as linear programming, PERT charts, simulation, etc., have been developed by the OR/MS community, but, due to their own success, have become public property. These tools are now used by many people who are not necessarily trained as OR/MS workers, and straightforward application of such tools is no longer the exclusive property of OR/MS. This coincides with Telgen's (1988) view of OR; he also notes that OR tools are increasingly being used as black boxes, e.g. while linear programming methods are included in spreadsheet packages, most users have no idea of the underlying algorithm. Development of fundamentally new tools and new ways of applying existing tools are necessary conditions to prevent OR/MS from eventually being sold out.

To conclude, there can be little question that, within the context of this book, OR should be considered primarily as a problem-oriented technology rather than as a knowledge-oriented science. This means that problem owners become a crucial element in OR studies, so it comes as no surprise that the relations between OR analysts and these problem owners have attracted a lot of attention; Section 17.2.2 focuses on the issues in client relations.

17.2.2 Client relations

We argued above that relations with clients are vital to applied OR. Here we look at three components of this issue:

- the frequently recurring theme of communication problems between OR analysts and their clients;
- the poor image of the discipline and the need for better public relations (PR for OR);
- the need for better understanding of how OR can be of value to clients, by the OR community and by clients.

17.2.2.1 *Communication between OR analysts and managers*

The problems that OR workers encounter in communicating results of their studies to managers were already recognized in the 1950s in *HBR*. Herrmann and Magee (1953) noted early that communication between OR workers and executives is the most serious problem with applied OR. It is interesting, though, to note the change in opinion in *HBR* as to who is responsible for breaking the communication barrier. Roy (1958) realizes that 'There is a vital need for a bridge between the two diverse frames of discourse represented by the prevailing loose terminology of the businessman and the precise language of the scientist', but stated that both business management and OR consultants should make an effort to overcome this gap. Bennion (1961, p. 100) gives reasons why 'more businessmen than not are highly skeptical of, if not downright antagonistic toward OR', but, in contrast to Roy, he believes that (p. 101) 'to correct this regrettable state of affairs, I am prone to feel that the econometrician and the programer [*sic*], rather than the businessman, must accept responsibility for the first step.' Jones (1966) already finds that communication from the OR side is deteriorating; writing about OR articles for businessmen, he states (p. 180):

> As a rough rule of thumb, the early articles tend to be better than later ones; the relationship between the business problem and the technique is frequently clearer in the former; and when they were written, there had been less time for a specialized jargon to develop.

In Chapter 18, we see that communication is still considered a serious problem, and that the first step towards bridging the gap will indeed have to come from the OR side.

That communication problems occur in both directions is remarked by Boulden and Buffa (1970, p. 66): 'Unfortunately, the OR staff often did not understand the problem, nor could the manager clearly define it.' An example of this is given in Chapter 18, where it only became clear several months into a project that the OR analyst had interpreted the problem differently than was intended. Although the communication issue does not seem to have greatly bothered the OR/MS community until relatively recently, it certainly does now. Smith and Culhan (1986) discuss the priorities indicated by teachers, researchers and practitioners of OR/MS: 'more than 50% of the 20 priority issues dealt with communications'. There is indeed a serious problem here. To non-OR people, there is inevitably something mystical about OR, which makes it harder for an OR analyst to communicate exactly what he or she is doing. The interdisciplinary nature of many project teams further exacerbates this problem, which becomes particularly acute if OR is to play a role in strategic decision making (see Section 17.2.4). The OR community seems to have entered the awareness phase concerning the communication issue, but it is not yet clear what can be done to help OR practitioners.

17.2.2.2 The image of OR/MS

The very first step in establishing relations with any client is made when the client invites an OR analyst to discuss a particular problem. But for this first step to be made, OR clearly needs to have a positive image among those prospective clients. However, practitioners can often be heard complaining that this image is not adequate. Compare the striking difference in tone between two *HBR* articles, 22 years apart. Baumol and Sevin (1957) began their article by writing 'It is difficult to exaggerate the opportunities for reduced marketing costs and increased marketing efficiency, and hence greater profits, which are offered to management by the combined techniques of distribution cost analysis and mathematical programing [*sic*].' On the other hand, recall Bishop's (1979) less euphoric view: 'The modus operandi of the operations researcher is to abstract from some real-world problem a mathematical problem for which he can find an answer. He then touts that answer as a guidepost to the manager.' The ongoing debate shows that the OR/MS community is aware of this problem, but is still struggling to find a solution.

One result has been to call the name of the field into question. The first paragraph of Wagner's (1975) well-known OR textbook discusses why the name operational research is ambiguous and unfortunate. Lilien's 'mid-life crisis' paper (1987) observes that

> the term *operations research* has developed a somewhat negative connotation in the late '60s and early '70s . . . the term *management science* is seen by many managers as a contradiction in terms; managers know that much of what they do is messy, ill-structured, based on poor or biased data, and in need of almost instant resolution.

As a possible remedy, he mentions that someone suggested using the term management engineering for the discipline of MS/OR. The Report of the UK Commission also recognizes the problem (1986, p. 854):

An extreme but widely held view – conceivably as many as half OR practitioners hold it – is that the phrase 'operational research' is now understood by clients in so narrow a way, and in a way so unrelated to the actuality of OR in practice, that some new name should be found.

The reason for not actually recommending a change of name is that the Commission could not agree upon a suitable replacement. And Weingartner (1987, p. 259) also is explicit about it: 'The view is growing in acceptance; the name of the field, and the name of the society that champions it, ought to be changed.'

Persistent management apathy toward OR/MS is one of the major trends noted by Geoffrion (1992), who claims, however, that managers are not to be blamed for this, as their main concern is to manage their own businesses. He does concede that OR/MS has a visibility problem, and that despite all the success stories available the business press largely has ignored OR/MS. Rinnooy Kan (1989) sees better visibility of the discipline as the solution to the image problem. Creating visibility is complicated by the interdisciplinary nature of much OR as it is difficult (if not unfair) to attribute a project's success to the OR component contained within it. And, of course, creating a better image requires better communication, another sore point. One important step can be made, though, by better understanding what it is that OR can do for its clients, and how that can be achieved.

17.2.2.3   How can OR/MS be of value to its clients?

Once the first step in establishing a relation with a client has been made, the main question becomes 'What can OR do for the client?' Although the OR/MS literature does by now contain an impressive number of excellent application papers, the process by which OR can be of value to clients is still poorly documented. Perhaps the most outstanding example of a collection of highly successful applications is given by the more than 20 special issues of the practice-oriented OR/MS journal *Interfaces*, containing the finalist papers for the Franz Edelman Award competition. These and many other unpublished success stories suggest that many practitioners are very skilful. However, most knowledge about the process of applying OR/MS is implicit, and generally seems to be learned by trial and error. For instance, many of the interesting points raised during the interviews on which Chapter 18 is based are rarely, if ever, discussed in OR textbooks. This situation is not very satisfying and even unacceptable for a mature discipline. The fairly recent book by Assad *et al.* (1992) constitutes a step towards correcting this situation, as does the current book.

One example of a simple framework attempting to capture the process of OR/MS is the OR cycle, suggested by Fortuin *et al.* (1992), which describes the phases a typical OR study would go through: from realization that there is a problem to problem description, model building and data gathering, selection of a solution method, validation, and, finally, implementation. Occurrences they warn against are solving the wrong problem (recall Boulden and Buffa's (1970) remarks about OR teams not understanding the problem, as illustrated in Chapter 18), and the fact that problems may change during the study, so that the implemented solution must be contrasted with the current problem, and, if necessary, the cycle must be repeated. In fact, in Chapter 18 several clients admit that at the beginning of a project, they sometimes do not know exactly what the problem is, so that they need the flexibility to be able to change the problem definition as the project proceeds. Combined with the potential communication problems mentioned before, this means that OR practitioners and their clients are not just aiming at a moving target, but doing so in the dusk!

Thinking about OR/MS in marketing terms may be helpful. Market segmentation is needed to distinguish between different types of clients (large vs small firms, private vs government, strategic vs operational, etc.) in order to cater for their needs. Key questions are how OR/MS can provide a value-added service to the client, and how OR/MS should evolve to gain and hold a competitive advantage over other concepts (the quality movement, just-in-time, business process re-engineering, etc.) managers could invest their scarce time and money in.

Above, in Section 17.2.2.1, we focused on the problems of communication. Particularly in the case of interdisciplinary projects, the contribution of OR to a project's success is often not (explicitly) recognized, which is one of the causes of the image problem of the field. The interdisciplinary character of OR will be the topic of the next section.

17.2.3   The interdisciplinary character of OR/MS

During its World War II origins, OR was conceived as a new way of solving problems (and therefore inevitably a problem-oriented technology), characterized by the interdisciplinary nature of the groups performing 'OR work'. Being at the heart of OR/MS, the question of interdisciplinarity has received much attention in the debate. A mathematical approach, whether this means using mathematical (but not necessarily sophisticated) techniques or simply an analytical style or logical thought, is an important characteristic of OR/MS, but it is precisely the dominance of mathematics over other disciplines that has attracted much criticism. From the following, it becomes clear that in particular the behavioural sciences are deemed to have been neglected by the OR/MS community.

Beged-Dov (1966) expressed his concern:

> Perhaps the single greatest obstacle to the establishment of OR as a powerful discipline noted for actual accomplishments is the fact that an ever increasing number of narrow specialists in mathematics and natural sciences impart to our profession undesirable dogmas and outlooks,

while Morse (1977) also observed a 'narrowing in outlook of many operations research workers'. Dando and Sharp (1978) are worried about the 'present difficulties OR workers have with problems involving major Social or Behavioural features'. The importance of interdisciplinarity is underlined by the UK Commission (1986):

> The Commission observed that much successful OR appears to be done in teams drawn from across an organization, seeded with one or more OR practitioners, who will play a key but not necessarily overtly central role in the structuring of the decision-forming process as the work progresses.

And Pierskalla adds (1987): 'If OR/MS is to grow, it must deal analytically and realistically with human behavior.' It must 'reach out to new areas of knowledge and to new approaches, and integrate them into our field'.

Chapters 15, 16 and 18 also illustrate that much OR work is interdisciplinary. Particularly when the decisions involved acquire a more strategic nature, incorporating the perspectives offered by several different disciplines is a *sine qua non*. Given that much successful OR work is performed in interdisciplinary teams, issues such as team management and communication between the various team members become highly relevant. In Chapter 18, several examples of how projects are managed are given.

17.2.4 OR/MS at the strategic level

It has been demonstrated beyond question that OR/MS can be highly useful in supporting tactical and operational decisions; whether or not OR/MS can be applied to strategic situations has been debated. Furthermore, a field that is not at ease with itself, that has problems communicating with executives, that is reputed to be predominantly mathematical in nature and tool oriented, will generally not be considered a useful aid in strategic situations. Many well-known examples, dating back to World War II, show the incorrectness of such a view of OR. In this volume, Chapters 11–16 all describe cases in which OR had a more or less strategic impact.

Although the name unfortunately suggests otherwise, OR should not be discarded too quickly when strategic issues are at hand. Salveson (1957) even stated that OR is primarily suitable for strategic decisions: 'Operations research has little, if any, role in making current operating decisions or actions. One sufficient reason is that, in the event of a current decision, there simply is not time for OR to help.' Hayes and Nolan (1974), however, take a different view; they believe that, while OR had proved useful in operating situations, trying to extend these smaller models to large corporate models led to 'disaster by addition' (p. 105).

Ackoff (1979a) observes that OR has been dispersed to lower levels within organizations. Weingartner (1987) is concerned that to have any impact on the higher levels within organizations, 'We must certainly avoid coming across as members of a priesthood, as practitioners of occult arts.' Rinnooy Kan (1989) also expresses his worries that OR ought to concentrate on achieving more strategic success: 'For instance, what will happen if tactical, short term planning procedures have been automated and organizations scrutinize their long term planning problems? Will our profession have anything of substance to offer them beyond what is available today?' Zipkin (1986) sees a clear challenge here: 'There are gaping holes in our collective abilities to model, especially at the level of strategy and policy, and therefore tremendous opportunities for energetic researchers and practitioners.' Kirkwood (1990) feels strongly about the importance of performing strategic projects:

> we should encourage those who are applying OR approaches to strategic problems. If the OR community does not recognize and support this work, then the practitioners doing the work will migrate to other communities. This will mean lost opportunities for OR professionals to work on interesting issues, and less visibility for operations research with corporate and government top management.

Three remarks are in order here. First, it may be true that, whether justifiably or not, managers do not see OR/MS as a way to help them when they are involved in strategic decision making. On the other hand, the implicit impacts of OR/MS on strategy, or, in Mintzberg's (1989) terms, its contributions to emergent strategy, are undeniable. The Edelman finalists again provide sufficient evidence that OR/MS can help shape a firm's strategy, whether or not that was the intention from the outset. Chapter 18, and the chapters in Parts 4 and 5, provide several examples of decisions with a long-term impact that were taken as a result of OR studies, albeit often as unintended results. Second, it is particularly in such strategic cases that OR/MS can only function and be taken seriously if it is part of an interdisciplinary effort, the subject of Section 17.2.3. This point is also repeated in Chapter 18. Finally, many papers in *HBR* have emphasized the importance of the learning effect of OR studies, which indicates that those authors believed that OR is a valuable support in strategic decision making; after all, for simple operational decisions, insight and understanding are perhaps less critical. For that reason, in the next section we turn our attention to the issue of learning from OR studies.

17.2.5 Learning from OR

Many *HBR* articles consider the learning effect of an OR study to be more important than the actual (numerical) results. The Report of the UK Commission contains some highly interesting observations (JORS 1986, pp. 841–2):

34. The Commission found significant amounts of O.R. in practice with one or more of the following aims:

- to help structure 'messes' or messy problems;
- to research into the facts of an uncertain topic;
- to help an understanding of a sphere of activity;
- . . .

35. It is noteworthy that the above list does not include, directly at least, optimizing operations or obtaining cost savings. While improving efficiency in some general sense is no doubt a driving force behind much O.R. in practice, it is achieved indirectly. Almost all practitioners with whom the Commission spoke think that the main benefits of O.R. stem from enhancing the client's understanding of his own problems. Practitioners commonly see their role as helping their client do his job better, not doing it for him.

Preparing for future decisions was seen by Salveson (1957) as the main use of OR. Baumol and Sevin state (1957): 'While a linear program usually will not compute a correct optimum because the changes it suggests will go too far, there is yet a very strong presumption that it will correctly indicate the best directions of change.' An article probing deeper into the interaction between OR and management (Roy, 1958), says: 'In short, and over-simplified, the OR method means getting behind the "art of doing business" and probing into what makes up that important but elusive thing called "business judgment".'

Bennion (1961; author's italics) is more explicit; he considered the most valuable aspects of models to be

(1) The capacity of the models to improve management's *understanding* of highly complex problems, which can scarcely fail to enhance the quality of the necessary value judgments management makes. (2) The undeniable fact that the most valuable use of such models usually lies less in turning out the answer in an uncertain world than in shedding light on how much difference an alteration in the assumptions and/or variables used would make in the answer yielded by the models.

Hayes (1969; author's italics) elaborated on this, and considered this the advantage of OR over other kinds of tools:

I believe that the greatest impact of the quantitative approach will *not* be in the area of problem *solving*, although it will have growing usefulness there. Its greatest impact will be on problem *formulation: the way managers think about their problems* . . . . In this sense, the results that 'quantitative people' have produced are beginning to contribute in a really significant way to the *art* of management.'

According to Brown (1970), the main benefits of decision theory analysis are not the numerical results, but the fact that it clarifies the relevant issues, makes implicit assumptions explicit, and provides a framework for communication. Boulden and Buffa (1970) concluded that 'Decision making is facilitated because the manager interacts with the model, not because decision-making logic is built into the model itself.' In Jones's words (1970):

> In business decision making, 'getting there is more than half the fun'. Thus many discussions of attempts to apply new mathematical methods to solving business problems end up with the conclusion that the real benefit was not the specific answer but the increased awareness of organizational problems and opportunities. This was achieved through the discipline of formulating the problem in a new vocabulary and structure.

Ackoff (1979b; author's italics) recognizes this fact: *'the principal benefit of planning comes from engaging in it'*. He suggests using the concept of 'idealized design', giving managers a direction to follow, even if the target itself may be unreachable.

Reading these comments, it is hard to escape the feeling that the OR/MS community has failed to recognize adequately how it could be of greatest value to managers, and it seems high time to develop in detail the implications of these remarks for applied OR. This is not to say that this learning effect is rarely achieved: Chapters 12–14 and other examples given in Chapter 18 clearly demonstrate the significant insights that can result from OR projects.

17.3 Conclusions

First, one could perhaps think of the sense of crisis as if it were caused by OR/MS reaching the next stage in the product life-cycle, gradually changing from a highly innovative approach to a more mature management tool. In doing so, its members are finding the boundaries of the field, inevitably leading to some disillusionment. However, this does not constitute failure, but is a consequence of success. Analogously, one might ask whether alchemists, despite having laid part of the basis of chemistry, failed just because they never managed to produce gold. There is no reason to believe that there is not a bright future for OR/MS, and every reason to believe that there is. It is up to the people in the field to bring that bright future nearer to home.

Second, some action is required to improve the visibility and image of OR/MS, e.g. by aggressively pushing suitable versions of the Edelman finalist papers into the business press. *Interfaces*, after all, is only read by the OR/MS community itself, not by potential clients. The field may not be paying enough attention to the business world; it certainly is not grabbing enough attention from it. The present book is one attempt at making the world of OR/MS more accessible to managers.

Third, it is important to note that nearly all papers referred to in this chapter are based on personal opinions or experiences of the authors concerned. Together, these provide valuable and rich insights into the practice of OR, but they should not be wrongly interpreted as being generally valid and proven statements. In fact, in Chapter 18 it becomes clear that the issues discussed here are rarely as clear cut as they are often made out to be in the literature. Perhaps the most important conclusion from this chapter and from Chapter 18 is that there is no such thing as 'the OR approach', but that the field of OR is flexible enough to let the approach depend on the problem, on the clients, on the environment, and on any other factor that might be relevant.

Finally, we see that, although much has been said and written, little is really *known* about the process of OR. If OR is to live up to its potential, which undeniably is enormous, we will need to investigate in much more detail the process by which this potential can be released.

References

ACKOFF, R. L. (1973) Science in the systems age: beyond IE, OR, and MS, *Operations Research*, **21**, 661–71.
  (1979a) The future of operational research is past, *Journal of the Operational Research Society*, **30**, 93–104.
  (1979b) Resurrecting the future of operational research, *Journal of the Operational Research Society*, **30**, 189–99.
  (1987) President's Symposium: OR, a post mortem, *Operations Research*, **35**, 471–4.
ASSAD, A. A., WASIL, E. A. and LILIEN, G. L. (1992) *Excellence in Management Science Practice* (A Readings Book), Englewood Cliffs, NJ: Prentice Hall.
BAUMOL, W. J. and SEVIN, C. H. (1957) Marketing costs and mathematical programming, *Harvard Business Review*, **35** (September–October), 52–60.
BEGED-DOV, A. G. (1966) Why only few operations researchers manage, *Management Science*, **12**, B580–91.
BENNION, E. G. (1961) Econometrics for management, *Harvard Business Review*, **39** (March–April), 100–12.
BISHOP, J. E. (1979) Integrating critical elements of production planning, *Harvard Business Review*, **57** (September–October), 154–60.
BLUMSTEIN, A. (1987) The current missionary role of OR/MS, *Operations Research*, **35**, 926–29.
BOULDEN, J. B. and BUFFA, E. S. (1970) Corporate models: on-line, real-time systems, *Harvard Business Review*, **48** (July–August), 65–83.
BROWN, R. V. (1970) Do managers find decision theory useful?, *Harvard Business Review*, **48** (May–June), 78–89.
CORBETT, C. J. and VAN WASSENHOVE, L. N. (1993) The natural drift: what happened to operations research, *Operations Research*, **41**, 625–40.
CORBETT, C. J., OVERMEER, W. J. A. M. and VAN WASSENHOVE, L. N. (1995) Strands of practice in OR (the practitioner's dilemma), *European Journal of Operational Research*, **87**, 484–99.
DANDO, M. R. and SHARP, R. G. (1978) Operational research in the U.K. in 1977:

the causes and consequences of a myth?, *Journal of the Operational Research Society*, **29**, 939–49.

DANIEL, D. W. (1987) Half a century of operational research in the RAF, *European Journal of Operational Research*, **31**, 271–5.

DRUCKER, P. F. (1959) Thinking ahead, *Harvard Business Review*, **37** (January–February), 25–152.

FORTUIN, L., VAN BEEK, P. and VAN WASSENHOVE, L. N. (1992) Operational research can do more for managers than they think!, *OR Insight*, **5**, 3–8.

GEOFFRION, A. M. (1992) Forces, trends and opportunities in MS/OR, *Operations Research*, **40**, 423–45.

HAYES, R. H. (1969) Qualitative insights from quantitative methods, *Harvard Business Review*, **47** (July–August), 108–19.

HAYES, R. H. and NOLAN, R. L. (1974) What kind of corporate modeling functions best?, *Harvard Business Review*, **52** (May–June), 102–11.

HERRMANN, C. C. and MAGEE, J. F. (1953) Operations research for management, *Harvard Business Review*, **31** (July–August), 100–12.

JONES, C. H. (1966) Applied math for the production manager, *Harvard Business Review*, **44** (September–October), 20–182.

(1970) At last: real computer power for decision makers, *Harvard Business Review*, **48**, (September–October), 75–89.

KING, W. R. (1967) On the nature and form of operations research, *Operations Research*, **15**, 1177–80.

KIRKWOOD, C. W. (1990) Does operations research address strategy? *Operations Research*, **38**, 747–51.

LILIEN, G. L. (1987) MS/OR: A mid-life crisis, *Interfaces*, **17** (March–April), 35–8.

MINTZBERG, H. (1989) *Mintzberg on Management: Inside Our Strange World of Organizations*, New York: The Free Press.

MORSE, P. M. (1977) ORSA twenty-five years later, *Operations Research*, **25**, 186–8.

PIERSKALLA, W. P. (1987) Creating growth in OR/MS (President's Symposium), *Operations Research*, **35**, 153–6.

Report of the UK Commission on the Future Practice of Operational Research (1986) *Journal of the Operational Research Society*, **37**, 829–86.

RINNOOY KAN, A. H. G. (1989) The future of operations research is bright, *European Journal of Operational Research*, **38**, 282–5.

ROY, H. J. H. (1958) Operations research in action, *Harvard Business Review*, **36** (September–October), 120–8.

SALVESON, M. E. (1957) High-speed operations research, *Harvard Business Review*, **35** (July–August), 89–97.

SMITH, R. D. and CULHAN, R. H. (1986) MS/OR academic and practitioner interactions: a promising new approach, *Interfaces*, **16** (September–October), 27–33.

TELGEN, J. (1988) Verzin een list! (in Dutch), Inaugural Lecture, University of Twente.

WAGNER, H. M. (1971) The ABC's of OR, *Operations Research*, **19**, 1259–81.

(1975) *Principles of Operations Research*, Englewood Cliffs, NJ: Prentice Hall.

WEINGARTNER, H. M. (1987) The changing character of management science, *OMEGA*, **15**, 257–62.

ZIPKIN, P. (1986) Confessions of an optimist, *Interfaces*, **16** (March–April), 86–92.

CHAPTER EIGHTEEN

What the cases don't tell us

C. J. CORBETT, W. J. A. M. OVERMEER and L. N. VAN WASSENHOVE

18.1 Introduction

When operational researchers write cases about their practice, they do so with the intent to show others what they have done and how they have done it, and to provide guidance for others on how to apply existing or new techniques to old or new problems. Cases are an important medium for the transfer of knowledge from one practitioner to another, or from seasoned practitioners to novices and students. Sometimes a case will also illustrate to a (potential) client how an operational researcher might be of assistance to him or her.

Over the past few decades, the OR community has realized that focusing only on mathematical techniques is not sufficient for becoming a successful practitioner. In the OR literature, much has been written about the dangers of focusing too much on theory (see e.g. Chapter 17 in this book or Corbett and Van Wassenhove (1993)). In response, the professional OR societies have launched several laudable initiatives[1] to emphasize OR practice more heavily, e.g. by providing descriptions of how such techniques are used in practical situations. Yet, if a novice to the practice of OR were to read all 120-odd papers on the highly successful projects chosen as finalists for the Edelman award for excellence in OR/MS practice, it is unlikely that that novice would be able to practise OR. In fact, in most cases, even a competent and experienced practitioner would have difficulties performing the projects as described and prescribed. Why is this?

We will argue in this chapter that successful practitioners do a lot more than they write about, and much of what they do not write about is in fact critical for their practice. Some issues may remain undiscussed because the practitioner is simply not aware of them. Very often, though, their omission seems due more to a pervasive form of what looks like a collective

Table 18.1 Overview of projects and people involved

Client organization	People interviewed	Consultant interviewed
Callebaut Europe (European industrial chocolate manufacturer)	F. Callebaut (director of operations)	L. Fortuin (CQM)

Project: developing and implementing performance measures

Monsanto Services International NV (multinational chemical company)	F. Cammaert (logistics manager Europe–Africa)	A. Van Looveren (Beyers & Partners)

Project: developing a system for supporting agricultural products sourcing decisions

Compco (multinational electronics company)	J. Smits (project leader) J. Van Gool (project team member)	H. Fleuren (CQM)

Project: simulation studies to aid in the design of a new plant for coloured liquid crystal displays

Oilco (multinational oil company)		D. Bookman (BOR Consultants)

Project: developing and implementing optimization software for cutting packaging foil

Transco (transport company)	W. P. Gray (head, vehicle maintenance department)	P. Brown (ABC Consultants)

Project: development of a decision support system for medium- and long-term labour planning decisions

RGD (Rijksgebouwendienst, government housing agency)	C. L. G. Wassenaar (research and development)	C. G. E. Boender (ORTEC)

Project: development of a system for decision support in allocation of housing to government agencies

AMEV Levensverzekering NV (insurance company)	D. A. Voûte (adjunct director) C. W. van Dedem (project team member)	C. G. E. Boender (ORTEC)

Project: development of a system for asset–liability management self-censorship, leading to cases that represent OR practice as a linear progression of activities invariably leading to a successful outcome. As a result, the ropes of practice remain hidden behind a streamlined simplicity suggested by many cases, rendering the learning process for others, particularly the uninitiated such as students, novices, and clients, unnecessarily difficult.

The editors of this book asked us to begin to address this situation by writing about the practice of three of the practitioners who contributed to this volume. In addition, we studied four more consultants. To break out of the potential hazard of limited representation by consultants of their own practice, we also interviewed members of the client system with whom the consultants worked closely together. An overview of the projects and people involved is given in Table 18.1.[2]

18.2 The limitations of prevailing views of the practice of OR

As our research progressed, we were struck by the two images of the practice of OR that seem to dominate the field. One is the perspective that OR practice is the mere application of OR theory, where the outcomes of (possibly black-box) computations are provided to the client who then makes decisions. This view provides a very narrow definition of OR, divorced from most concerns of practice; the limits of this image are widely recognized within the OR community. The other image, which one might infer from reading OR case descriptions, is that OR practice involves a linear process of problem solving within a limited amount of time, proceeding by some well-defined sequentially linked stages such as problem definition, model building, data gathering, model solving, validation and implementation (see, for instance, Fortuin et al., 1992). This image, too, is limited. It does not capture the fact that practitioners are not always immediately successful in a first attempt to address a problem, that trying to solve a problem can lead to new information that throws a new light on the original problem, that problems sometimes have to be reframed as a result, that projects get stuck and sometimes unstuck, that minor and even major disasters occur during projects, and that practitioners respond to and often learn from all these events. These aspects of practice increase the complexity of projects in non-trivial ways, which are mostly unaccounted for in theory and in accounts of practice, even though these are key ingredients of competent practice of the case writers.

The project came about when one of us, Charles Corbett, educated in a more or less traditional academic OR department and knowledgeable about OR techniques, joined a business school PhD programme and began to wonder exactly how such techniques were used in practice, and what practitioners actually do besides straightforward application. Together with Luk Van Wassenhove, a professor of OR with a concern for use of OR in practice, and Wim Overmeer, a professor in management focusing on strategies for organizational learning and enquiry, he decided to find out by talking to highly competent OR practitioners and map out how projects are conducted, not only based on the account of the practitioner but also on that of the user–client.

The observation that there are many important, non-technical aspects that have an impact on the outcomes of OR projects is by no means new. Ever since the early days of OR, people have warned that communication problems between client and consultant occur frequently, and that all sorts of political issues within the client organization can play a role. However, these issues are rarely discussed in any depth in a case description or elsewhere, essentially conveying to readers either the somewhat fatalistic message that 'things can and will go wrong, but there's nothing you can do about that', or relegating the problem to 'organizational behaviour'. In this chapter, we intend to describe aspects of practice that cannot be ignored, and that will hopefully eventually lead to a new image of OR practice. Such an image is, in our view, essential, to mitigate the often slow, painful and sometimes damaging learning process of beginning OR practitioners. Although just a first attempt, we hope it will constitute a step towards answering the question: 'how can we educate someone to become a successful practitioner?' It might also begin to provide the more experienced practitioners with a way to understand better their own competences, and help them reflect on their own effectiveness.

We will begin by offering brief descriptions of the seven projects we studied and highlight critical aspects of the process followed by the practitioners in collaboration with their clients. In the second part of the chapter we will turn our attention to three key problem areas facing practitioners as well as concerned academics:

1 *Project management issues.* We will focus on the two related issues of problem framing and project planning. The view of OR as an application of techniques assumes that the client has set the problem and that the task of the OR practitioner is merely to select the best techniques given the way the problem is formulated. The image of OR as a problem-solving cycle assumes a sequential process of defining, analysing and solving a problem, and then implementing the solution. However, we found that the problem need not be framed by the client, or that the framing of the problem shifts as the project takes place. Moreover, some consultants and clients start with just a hunch, or a very broad problem definition, while in other cases, they begin with a narrow problem definition. We will return to these issues in Section 18.4.

2 *The interaction between client and consultant.* With differing ways of framing problems and planning projects, the interaction between client and consultant takes centre stage. Key actors in the system – those who take the final decision, those who frame the problem, those who provide input, those who use the outcomes – tend to change, and this can cause

significant disruptions to the project. In such a context, the gathering of valid data is a non-trivial and often underestimated problem. Moreover, communication problems between client and consultant are likely to arise either during the framing, the implementation or the reframing. Finally, the meshing of the client's expertise about his or her own problems and that of the consultant about OR can be complicated: some (but not all) clients have considerable technical, mathematical or even specific OR expertise, while some consultants (but again not all) have gained substantial know-how about the client's domain through past projects. These issues are addressed in Section 18.5.

3   *The learning effect.* This is frequently referred to as an important if not the main benefit of an OR project. However, it remains an elusive and even somewhat mystical effect, and frequently occurs as an unexpected byproduct rather than as the primary goal of a project. What exactly does a client learn, and how? This issue will be addressed in Section 18.6.

In the rest of the chapter, our main purpose is to illustrate that the images of OR as an application of technique, and of OR as a linear problem-solving cycle, are seriously incomplete overall prescriptions for OR practice. They provide a false sense of stability, a stability we have not found during our discussions with successful practitioners. The framing and reframing of problems, the sometimes open-ended projects, changes in the client system, the meshing of complementary and overlapping expertise, problems in data gathering and communication, all lead to what we believe is a more realistic image of practice that is messier, more unstable and more ambiguous. Under those conditions of practice, practitioners display skills not captured in the traditional images of OR. Yet, we believe that those skills are central to competent and successful practice. Learning to be an effective OR practitioner hinges on acquiring those skills, which constitute the ropes of OR. In this chapter we point out some of the key areas in which skills need to be developed. In a follow-up paper (see Corbett *et al.*, 1995) we have tried to develop ideas about how such skills are acquired.

The rest of this chapter is structured as follows. First, in Section 18.3, we provide preliminary outlines of the seven cases on which this chapter is based. The outlines focus entirely on the non-technical aspects of each case, and in particular on the questions raised above. In Sections 18.4, 18.5 and 18.6 we discuss each of those questions using the evidence provided by the cases. In the conclusions, we place the findings of this chapter in a broader context, and suggest how they relate to a number of concerns held by many within the OR community.

18.3   The cases

We now present seven outlines of cases that illustrate critical features of the process of OR. The first two cases – 'creating meaningful performance measures' and 'optimizing global sourcing' – are two small projects by two different consultants. In both cases, the problem was relatively clearly defined by the client. The next two cases – 'simulating a complex production process' and 'implementing a customized cutting stock system' – are larger projects. In both cases, the client had a fairly clear sense of the problem. However, in the first case, the consultant was 'on call', doing a series of smaller projects and two larger ones. In the second case, the client asked for two pilot studies before embarking on the project itself. In the fifth project, on 'decision support for tactical labour planning', the consultant was asked to identify promising projects. The last two cases – 'decision support for office allocation' and 'jointly developing a decision support system for asset–liability management' – are different in that the client had a broad and open problem definition, and then engaged with the consultant in a process of narrowing and reframing the problem definition.

18.3.1   Callebaut: creating meaningful performance measures

18.3.1.1   *The client framing the problem*

Frans Callebaut, director of operations at the Belgian chocolate producer Callebaut Europe, had been implementing various improvement programmes throughout his plant. But he became aware that it was difficult or impossible to say whether these programmes were really working. He concluded that he needed a performance measurement system. Callebaut had a clear sense of what he wanted – the formulation and implementation of performance measures that would be accepted by his employees as reasonable and practical throughout his operations division. But he did not know how to do this.

18.3.1.2   *The consultant improvising on previous experience*

Leonard Fortuin, at the time with CQM, formerly an internal Philips staff consulting group, had been asked by a Philips director to write a report on the use of performance measures within Philips. When, some time later, Fortuin was interviewed by an engineering journal, about performance measures, Callebaut read that interview just as he was looking for someone to help him introduce performance measures in his division. He invited Fortuin for a first meeting to discuss the situation. Because Fortuin had no experience in chocolate production, he subsequently visited the plant, and let himself be guided around by employees to make sure he understood it well enough to be able to conduct the project confidently. From the many projects he had performed within Philips and CQM, Fortuin had accumulated a lot of experience related to production processes. This gave him an intuitive feel for the type of measures that would be needed in Callebaut's case, and how they could be implemented in practice.

18.3.1.3 Creating meaningful measures

Fortuin developed an action plan, which he discussed with Callebaut in a second meeting. They decided to let functional work groups of employees define their own performance measures, in brainstorming sessions, supervised by Callebaut and Fortuin. This way, Callebaut could be sure that the people who would have to implement and use the performance measures would support them and that the measures were practical.

18.3.1.4 The consultant as facilitator

The action plan was carried out in a one-week implementation session. During that week, Fortuin acted as facilitator in the group sessions. Callebaut was present all the time, so whenever a measure was agreed upon, he could authorize it directly; when an employee would remark that such data were hard to get, Callebaut would instantly order that they find out why those data were hard to get, and how they could be collected in the future.

18.3.1.5 Extending the performance measurement system

After about three months, enough data had been collected to make the improvements resulting from Callebaut's efforts visible; this was as Fortuin and Callebaut expected. On reviewing this chapter, Callebaut added that 'the performance measurement system is working very well and the key performance indicators are now used during the Board of Management meetings as a base for further actions', and that 'a similar performance measurement system has been implemented in the meantime as well at our subsidiary in the UK'. He lamented that it is very hard to push the performance measurement system beyond the operations division.

18.3.2 Monsanto: optimizing global sourcing

18.3.2.1 The client sensing an opportunity and preparing himself

François Cammaert was the head of the logistics department of Monsanto for Europe and Africa. Together with a financial analyst, he was thinking about the issue of global sourcing of agricultural products: for each customer, from which plant worldwide should the raw materials be sourced, where should the product be formulated and where should it finally be packaged? A trade-off had to be made between production costs, transport costs, and import duties and profit taxes, each of which depended on the routings chosen. Cammaert was convinced that Monsanto could do better than its current unstructured approach to this problem. He and his staff had already spent two years thinking about how to improve their product sourcing decisions. He had even formulated a simple mathematical model describing the problem. What he needed was technical expertise to assist in developing and implementing a working version of his crude initial model.

18.3.2.2 The client's problem formulation improved by a consultant entering a new domain

When Cammaert attended a seminar on linear programming organized by the Belgian OR society in which Beyers & Partners took part, he was struck by their emphasis on applications of LP, so he invited them to come to Monsanto. Beyers & Partners had no particular experience in global sourcing problems, but they were experts in applications of LP. During their first meeting they could instantly point out the shortcomings in Cammaert's own model and suggest a modular approach, which was exactly what was needed.

18.3.2.3 Little formal planning and quick results with a prototype for part of the problem

The Monsanto project was a small one, encompassing a period of less than two months, during which time it was not a full-time concern for either of the two Monsanto staff or for Anita Van Looveren, the consultant involved. No formal project planning was performed, although clear agreements were made on project content, time involved, budget, and delivery date, for both of the two-month stages: (1) modelling and validation, and (2) application and documentation. One of the difficulties Cammaert encountered was in obtaining figures for import duties and tax rates. The fact that Cammaert had already devoted so much thought to the problem meant that he had a very deep understanding of the issues involved, and could quickly explain any surprises that occurred in building and using the model. After four or five weeks, the first prototype was completed. Even though only part of Monsanto's operations had been included, it already gave useful results. The remaining four weeks of the project consisted largely of improving what had already been done, such as improving output screens, and of documenting the model.

18.3.2.4 Results of the project, including a new project

The model has been used several times, to choose routings given the current situation, and also to evaluate possible changes in the network. It is a strategic model, and is run approximately once or twice a year, whenever the situation has changed sufficiently to merit a revision. It has already saved at least $500 000 to $1 million, and has given Cammaert and his staff various useful insights into the structure of the global sourcing problem. Although this was Cammaert's first encounter with Beyers & Partners, he came to trust them enough to engage them in a second, much larger project, even though Monsanto headquarters in the United States advocated a different consulting firm, taking a certain personal risk in doing so. On reviewing this chapter some time after the interview, Cammaert wrote that 'the evidence after another 2 years is however that the system is less used than in the past, since several of the constraints, which were at the base of the model, have been designed away since then. Maybe this is another hidden benefit of this modelling approach!'

18.3.3   Compco: simulating a complex production process

18.3.3.1   *A client with a hunch, and an atypical project for the consultant*

In 1988, Jan Smits became leader of a project team within Compco, responsible for designing a plant to produce coloured liquid crystal displays, with many consumer electronics and industrial applications.[3] This was to be the first such plant in Europe, so no experience with the production process involved was available. The project team consisted largely of engineers, each responsible for part of the total production line, which comprised several hundred process steps. In designing the line, the team had to trade off factors such as throughput times, work in process and capacity, but had little insight into how these factors related to one another.

While Smits had no previous experience with OR, he had, as a civil engineer earlier in his career, been involved in determining the tensions in ram and pile, using simulation; as a result, he recognized that the static spreadsheet calculations already performed in designing the plant gave insufficient insight into what were inherently dynamic processes. 'Piling isn't a static process either, it's dynamic, you're hitting something, and hitting is never static. It was a very different simulation programme, but that explains my enthusiasm to find out quickly how [the plant] would behave dynamically.'

18.3.3.2   *A pilot simulation of a subprocess*

Smits asked CQM, a consulting group which had close links with Compco, to perform some simulation studies. Fleuren, the consultant, could contribute both his simulation expertise and his experience in modelling production processes from previous projects at CQM. Smits first commissioned a simulation of one of the subprocesses, by way of a pilot study, to see whether he and Fleuren could get along well. Smits was sufficiently satisfied with the results to ask him, in 1989, to perform a simulation of the entire line, which took Fleuren about 50 days. To use the words of Smits, the aim of the simulations was to know more about key design parameters of the plant: 'What should I pay attention to? If I turn this knob, how will the plant react?' Fleuren was then called in another three or four times for various small projects on subprocesses. Fleuren remarked that Smits and he had an efficient way of framing the problems: Smits would call Fleuren to talk about some subprocess that needed detailed analysis, after which Fleuren would write a proposal describing the problem, the approach he had in mind, and the time the analysis would take. That way, Smits could check whether Fleuren had interpreted the problem correctly.

18.3.3.3 Problems with data gathering, and finding an intermediary to help out

In 1992, when the design of the plant was nearing its completion, Fleuren was asked to do another simulation of the entire line. Particularly in the early stages of Fleuren's involvement, the design of the as-yet non-existent production line continually changed, but adapting the simulation model to each new design would have been prohibitively expensive, so a decision had to be made which configuration of the line to simulate. In the overall design project, certain decision points had been specified; the configuration as determined at one of these decision points was chosen for the simulation. Fleuren pointed out that this was not a typical project: it is more common to be called in to study an existing production line rather than to be involved in the design stage. The problems they are then confronted with could often have been prevented with more careful design.

Gathering data was often a problem. To be able to perform useful simulations, Fleuren needed accurate estimates of machine breakdown frequencies, processing rates, etc. Some of the process engineers did not see the importance of such figures, and could not be bothered to come up with precise estimates. Fortunately, one project team member, Van Gool, helped make Fleuren's task easier. Being a member of the team (which Fleuren was not), Van Gool was closer to the process engineers, and knew how to get the data needed.

18.3.3.4 What the project team learnt from the simulations

The simulations led to a number of important insights for the project team. They brought out, among others, the drawbacks of a line which is too carefully balanced, the importance of the location of the maintenance staff, and the effects of different material handling systems. This resulted in various minor and major design changes in the production line, which represents a total investment of several hundred million dollars.

18.3.4 Oilco: implementing a customized cutting stock system

18.3.4.1 A long incubation

Two years before the project actually started, some staff members at Oilco in Belgium believed they needed customized software for their packaging foil cutting machines. BOR Consultants was already acquiring a reputation for their experience with cutting problems, a reputation which has grown since. The Oilco staff visited BOR, went back to talk it over, and returned a year later. There was a more primitive system in use at Oilco in the United States, but the Americans were persuaded not to impose it on Europe too, so eventually permission to do a project with BOR was granted; the project was performed by a team, with David Bookman as consultant in charge.

18.3.4.2 *Two pilot studies, and tight project planning*

The project started with two pilot studies: (1) a functional pilot study, concerned with defining the problem in detail, and specifying what the software should be capable of, and (2) a more organizational study of hardware and software aspects, revolving around issues such as how to build the interfaces, who should install which database, on what hardware, what type of output the system should generate, etc. A presentation to general management was made after the first preliminary study. For the main body of the project, a contract was drawn up that distinguished a number of stages. Although much of the software had to be developed from scratch, BOR were able to plan and execute the 18-month project very smoothly and tightly, based on their previous experience in this domain. Despite some changes in the composition of the project team over its 18-month total duration, deviations from the plan were limited to a few weeks. Integrating newcomers into the team provided no serious problems.

18.3.4.3 *Coaching the users*

Cooperation with the planners, the future users of the cutting software, generally went smoothly. However, given the size of the plant and the cutting machines, it is understandable that planners sometimes felt uneasy about fully transferring control to the new software. In such cases, it was Bookman, the consultant, who had to coach them along, e.g. by being present during the first days after the actual changeover to the new system, and by installing a modem link with the BOR office. As Bookman summarized his coaching of the planners: 'I'll jump first, you just follow me'.

18.3.4.4 *A new project*

This project concerned the optimization of an existing process; as a result of its successful completion, BOR were subsequently called on in a project involving the actual selection of machines.

18.3.5 Transco: decision support for tactical labour planning

18.3.5.1 *A committed client*

William Gray had just been appointed head of the 2500-strong Vehicle Maintenance Department at Transco, a very large transportation firm. He has an engineering background, and tries to apply analytical reasoning wherever he can. He did not have any particular problem in mind that urgently needed solving, but held a strong conviction that OR could help him run his division better. The in-house OR group at Transco had suggested to Peter Brown of ABC Consultants to get in touch with Gray. The latter invited Brown to talk to the heads of each of the subdivisions and draw up a list of possible projects.

18.3.5.2 *The consultant defining projects*

It was left to the consultants to come up with good proposals, which they were able to do, despite not being familiar with the problem domain. These were presented to Gray, and they started on the one promising the largest return. This first project was to develop a system to help Gray determine how much staff he would need for a given type of activity in his division, for given work schedules, for given collective labour agreements, for given organization structures, etc. The system would help with a range of medium- and long-term labour planning issues, such as negotiating labour agreements.

18.3.5.3 *Working in an informal way*

Project planning had not been done in a very formal and precise way. This is precisely how Gray wanted it. He remembers that 'that project was somewhat untransparent, but if we have a good relationship with a consultant we don't make very formal fixed contracts'. This fits well with Brown's view: 'We usually don't build systems the way a computer scientist would advise, i.e. defining upfront what the system should do and adding increasingly more details.'

18.3.5.4 *Changes in the client system, and reframing of the project*

Unfortunately, by the time the system was completed, Gray had moved to a different division within Transco. His successor was not interested in the system, and Brown was relegated several levels downward in the hierarchy. However, for the people at that level, the type of problem Gray had been concerned with did not exist; for them, labour agreements etc. were a given, and considered unchangeable. They do use the system, but for purposes of a far more operational nature than intended, such as determining how much work the division will have during the coming period, or determining detailed work schedules. In the original system, the detailed work schedules were not relevant in themselves, only as a means for evaluating the impact of higher-level decisions. The system has never been used for its original purpose. Gray, however, is still very enthusiastic about OR, and since having moved to a different department, he has initiated several projects there with Brown and ABC.

18.3.6 RGD: decision support for office space allocation

18.3.6.1 A client senses an opportunity

RGD is a Dutch government agency in charge of housing for all government agencies. The director of the research department of RGD had been struggling for some time with the problem of allocating buildings to organizations. When he happened to see a personnel advertisement from ORTEC he realized, from the way ORTEC described itself, that they were the kind of consultants he needed. As a result, ORTEC and he set up a project, called 'BOSS', to develop a system for the building allocation problem; for ORTEC, this was an entirely new field. While that project was under way, the Ministry of Health and Environment (VROM) announced a move to a new building. Such a large-scale move often entails employees having to change office several times. Kees Wassenaar, a member of the research department and an expert in the realm of housing for government agencies, wanted to help minimize the number of such internal office changes that would be needed during the move. He knew of BOSS, so he contacted Guus Boender of ORTEC. Together they initiated a new project, which was 'strongly aimed at reducing the number of times people had to change offices during an internal move within one building', as Wassenaar put it.

18.3.6.2 Client and consultant favour informal cooperation

The intention was to complete the project in time to be able to assist in planning the upcoming move. But, rather than draw up a contract and a detailed work programme for the entire project, Wassenaar and ORTEC decided roughly which direction to take and worked on the basis of periodically renewed fixed-term, fixed-sum agreements. There was no clear goal to be met in each period, both parties proceeded on the basis of mutual trust. Every time Wassenaar and a colleague discussed the problem with ORTEC, the consultant would write a brief report to test their understanding of the problem with Wassenaar. After three months a proposal resulted, with a suggestion for an approach and an estimate of costs.

18.3.6.3   *Underestimating complexity leads to reframing the project*

The original goal turned out to be much more complex than either side had foreseen, and they were not ready by the time VROM moved into its new building. In response, they decided to simplify the problem. The focus of the project evolved from allocating agencies to buildings into allocating office space within a building to departments of agencies. Wassenaar stated:

> We decided to concentrate on simply matching organizations and buildings, before trying to do something about moving organizations around. We couldn't even find a good matching within one building, so how can you even think about reducing the number of moves in going from one matching to another?'

Allocation of office space within an agency used to depend on which department had the most power; Wassenaar hoped that with a system like this the allocation procedure could be made much more objective.

At the same time, Wassenaar recognized the need to identify explicitly a new target group, to keep the project moving, because he felt that 'there has to be a deadline, then people try harder'. With the change of focus, the group of potential future users of the system also changed from site managers wanting to reduce the number of office changes to housing consultants within RGD, who check whether certain buildings are suitable for certain organizations. For instance, if an organization has five buildings to choose from, the consultant has to determine how well the organization can be accommodated in each of the buildings.

18.3.6.4   *The original approach appears based on a false assumption*

One of the major complications in the early stages of the project was that the pilot study was based on what turned out to be a false assumption. The approach proposed was to generate a large number of admissible allocations meeting all relevant criteria and then choose the best of these allocations. However, this presumes that such admissible allocations exist! It turned out that no allocation met all criteria (hence, perhaps, the traditional power struggles), so the problem had to be reframed as finding a collection of allocations that would meet as many criteria as possible, and would let the final decision makers choose from that collection, based on which criteria they perceived as more important.

18.3.6.5   *An unintended further shift in problem focus*

A natural approach in tackling the problem of allocating office space to entire departments is to determine an allocation of individual people to individual offices. As Wassenaar said, 'the concept of allocating people to offices became very dominant, so the software's output was in terms of people and offices, but that isn't really relevant. What matters is the differentiation of office space.' Although both parties seemed to understand the problem, the project unintentionally shifted from its original, more global goal to the more detailed level. It was only when Wassenaar decided to stop working on the algorithms for a moment and focus on the output screens that this shift in problem framing became apparent.

18.3.6.6   *The client contributes to the algorithmic development*

Boender believes that the client's contribution to the project was critical for its success so far: 'the people with whom we dealt had an exceptionally deep knowledge of the issues they were facing and grasped our approach and our contribution quite quickly'. Wassenaar confirmed this when he said

> At some point in time you want to know how the algorithm works, you think they've understood the problem but when you see the algorithm you see that everything works in one particular way and you wonder, why not some other way?

Hence, it took ORTEC's experience in developing solution methods combined with Wassenaar's insight into the problem to construct an efficient solution method.

18.3.7   AMEV: jointly developing a decision support system for asset–liability management

18.3.7.1   *The consultant senses an opportunity to expand on previous work*

The history of the development of the asset–liability management (ALM) system started around 1984, when the Rabobank hired Guus Boender of ORTEC to help it develop a system for liability management for the Rabobank's own pension fund. When the project was completed, Boender saw that if the system were expanded to include a wider variety of assets, it would have great potential. Because of changes in the legal and economic environment, pension fund managers needed a new type of tool to evaluate investment strategies.

18.3.7.2   *The consultant sets up joint ventures with clients to develop the system further*

Together with three partners, one of which was AMEV, a major insurance and pension fund management company, Boender further developed the ALM system.[4] The connection with AMEV was made through a former colleague of Boender's who then worked in the investment department at AMEV. This former colleague persuaded AMEV's Voûte, then in the liability department, to cooperate with ORTEC. Voûte saw the possibilities of the consultant's system but also identified serious shortcomings for the applications he had in mind. To develop the system further in AMEV, Voûte set up a project team, consisting of a business economist (himself), a macroeconomist, a lawyer (Coen van Dedem), some actuarians and an econometrician; most of the members of the project team at AMEV knew nothing about OR. The project started on the basis of what Boender had developed for the Rabobank. The intention was to combine Boender's knowledge of OR with AMEV's knowledge of pension fund management in order to develop a more powerful system for ALM. ORTEC intended to sell the system to various pension funds in the future, AMEV was looking for a tool to help it assess the impact of different investment strategies and pension schemes, as part of the consulting service it offered to small pension funds.

18.3.7.3 From very loose to very tight project planning

Initially there was no project planning at all; the project started as a very loosely organized joint venture. ORTEC was not operating as a consultant to AMEV, and several AMEV staff invested a large amount of their own time in it. During the first serious meeting, the team estimated it would need about half a year for the project. Eventually, 'it took half a year or more just to understand each other', as team member Van Dedem puts it.

Choices as to what to include and what to leave out, such as whether to include foreign currency issues etc., were continually being made throughout the project. Initially, the team became more and more ambitious, and wanted to incorporate more and more, but at a given point in time, Voûte said, it 'came back to Earth, with both feet on the ground', and started eliminating options. For example, the team initially tried to capture career paths of individuals within a firm in a very precise way in the system, but then realized that this would be far too complicated. ORTEC suggested a much simpler, approximate method, which was then used. The framing of the problem did not change in any fundamental way during this project, but it gradually became more specific.

It was Voûte who decided, in November 1991, well into the project, that some more time pressure was needed. After taking stock of the work that still had to be done, a deadline was set for May 1992. In fact, Voûte announced that a presentation about the project would be given, and invited a large number of people from outside the team to set a firm deadline. From that moment on, everything was planned very tightly, tasks were assigned week by week, people reported back and forth; they slipped from that plan by less than a month.

18.3.7.4 The client reorganizes as a result of the project

For AMEV, the project had major consequences. Traditionally, the firm had an investment department and a pension liability department, so each pension fund had to deal with people from two departments within AMEV. The ALM system formally unites the two perspectives, which had previously been operating separately. Correspondingly, AMEV reorganized its pension fund management service, so that they now have one special department dedicated to supporting pension funds, of which Voûte is director; most members of the project team now also work in that new department. What ORTEC now has is a 'system as basis for customization'. With the larger pension funds, there are enough actuaries and econometricians with whom ORTEC can perform studies; the smaller funds are serviced by ORTEC's partners.

18.3.7.5 *Enthusiasm carries the project along*

Voûte, without prior experience with OR, found that it was initially not easy to understand the OR consultants. For those involved, most of the work had to be done in addition to their normal workload. Owing to a communication gap that existed initially, it took the team more than six months to reach a 'common frame of reference'. A key success factor was the enthusiasm of the project team, particularly given the unofficial nature of the project. Van Dedem, one of the members, remarked: 'We all found the subject so interesting that we all got carried away, we all became as enthusiastic as Guus [Boender].' They perceived the project as an intellectual challenge.

Having sketched the seven cases, it is now possible to look into several critical issues such as project management, client–consultant system issues, and the learning effects related to each of the cases.

18.4 Project management issues

18.4.1 Issues

With respect to project management, two issues seem particularly important: that of deciding what the problem is, and that of planning the project. The two are obviously related, but we discuss them separately here.

18.4.1.1 *Who frames the problem?*

While the prevailing images of OR practice assume that a problem has been framed early on, and that it has been fixed, we found much more variety in this area. In the Transco case, we saw that the consultant was invited to look for problems and define them himself, whereas in the Callebaut and Monsanto cases the respective clients followed the opposite approach by framing the problem before calling in a consultant. Callebaut had done some preparatory work before inviting Fortuin, and believes that when you engage consultants, it's very important that you've written down a rough draft of what's expected. I believe that's essential, that you make the consultant's task very clear from the beginning. You have a sort of proposal, about three pages, then you can see how he reacts to it.

Similarly, it was Monsanto's Cammaert who wrote down a brief description of what he wanted from the consultants: 'I think that's what we should do, we should write down all we know and say "look, this is how far we've got in our thought process".' In Cammaert's opinion, such a degree of preparation on the client's part is necessary before embarking on an OR project.

18.4.1.2 *When is the problem framed?*

The problem as framed by Cammaert was very precisely defined: 'To be able to apply OR you first need a basis. Not all problems are sufficiently well developed to bring them within the reach of OR.' In the Compco case too, Fleuren and Smits made sure that they agreed on what to study. Alternatively, the framing of the problem is not necessarily completed at the beginning of a project. The ALM project, for instance, started without an unambiguous definition of what the aim of the project was. The framing of the problem did not change in any fundamental way during this project, but it gradually became more specific. Van Dedem said: 'Some parts of the system became more detailed, of others we decided not to include them in a detailed way but in a rougher way instead.'

18.4.1.3 *Shifts in problem framing?*

Lastly, it became clear that a problem, once framed, is not immune to shifts, which can be caused by a variety of reasons. The focus of the Transco project shifted to a far more operational level when the project was delegated to a lower level in the hierarchy, as a result of Gray's departure to a different department. In one of the cases, although agreement had been reached on what the project would involve, one of the clients changed his mind halfway through: suddenly he wanted the project to embrace the total production control system rather than the one particular issue originally singled out. This would of course have required a complete revision of the project, so the consultant could not and did not agree; the project remained as initially intended. Sometimes a shift in focus is desirable, though: in Bookman's experience, clients often come with some operational manifestation of a problem, whereas the actual cause is often at a deeper, more strategic level.

The RGD project witnessed three shifts. First, when it became clear that the project would not be completed in time to assist with the Department of Environment's move, Wassenaar and Boender decided to set themselves a less ambitious target. Second, when the development of algorithms was in full swing, the focus shifted unnoticed from allocating office space to departments, to allocating individual people to individual offices. It was only after the decision to stop working on the algorithms for a moment and focus on the output screens that this shift in problem framing became apparent. Third, when admissible allocations turned out to be non-existent, the objective had to be reframed as finding a collection of allocations meeting as many criteria as possible, and letting the final decision maker choose from that collection, based on which criteria were perceived as more important.

18.4.2 Project planning

18.4.2.1 Tight or loose planning?

Whether a project should be tightly and precisely planned does not seem to depend on the size of the project. Each of the simulations performed by Fleuren constituted a relatively small project, but for each of them he would write a proposal describing the problem, the approach he had in mind, and the time the analysis would take. The Monsanto project was also a small one, but here little formal project planning was performed. 'Let's just begin, we'll see where we end up', as Cammaert said. He sees no advantage in formally planning such a small project, as so few risks are involved; in fact, he does not really believe it is possible, 'you would be introducing a handicap'. The only formal planning that was done was intended more to pacify the purchasing department, which controlled all expenditures.

18.4.2.2 Early or later?

We find the same variation in large projects, as in the Oilco and ALM cases. The Oilco project was planned in detail, with two pilot studies, and executed accordingly. According to Bookman, distinguishing well-defined stages in a project contract has three reasons: first, to monitor progress; second, to define checkpoints at which tests are to be performed or documents handed over, and to link payments with these; and third, the client also needs to write certain interfaces, so the consultants need to know when certain parts of the software will be available to them for testing. In the ALM project, there initially was no project planning at all. Voûte decided, well into the project, that more time pressure was needed, and from that moment on, everything was planned very tightly.

18.4.2.3 Reducing the impact of changes in the client organization

Although precise planning was perceived as undesirable by several consultants and clients, it does seem to be a way of reducing the impact of changes in the client organization. Compare the Oilco project where, despite some changes in the project team over its 18-month total duration, deviations from the plan were limited to a few weeks, to the Transco case, where Gray's departure led to a drastic change in focus of the project and to the abandonment of the other proposals which Brown and Gray had informally agreed on. Gray himself, however, is in favour of a flexible approach. About ABC, he says 'they're flexible, informal; they don't keep on nagging about what's formally in the contract; that's good, because as client you don't always know exactly what you want, other things sometimes turn up unexpectedly'.

18.5   Client–consultant system issues

18.5.1   Changes in the client–consultant system

18.5.1.1   Effects of changes

Especially when dealing with larger organizations and with longer projects, consultants have to be prepared for changes in the people they deal with. Several consultants remarked that such changes have become a fact of life in long-term projects. In Bookman's experience, it is normal 'that all the people with whom you made the initial agreements are no longer there when it's finished. That can be very disturbing, that's why a good pilot study is so enormously important.' Van Looveren expressed similar experiences. Fleuren, too, had to convince each newcomer of the value of his simulations every time the composition of the design team changed. Wassenaar, who conducted the office space allocation project with one colleague, admits that 'if either of us had been given a different function, the project would probably have fizzled out'. We have seen the effects of Gray's departure to a different department within Transco. An open question, though, is whether a more formal approach would have prevented Gray's successor from relegating the project downwards in the hierarchy. People changing place during projects is a common occurrence, Brown says, and to make oneself less sensitive to such changes one might have to accept the 'necessary evil of defining much more clearly what you're doing'.

18.5.1.2   Coaching newcomers

Clearly, consultants need to spend a lot of effort on coaching newcomers into ongoing projects, to ensure that the newcomers feel equally involved as their predecessors. But even after a project is completed and, for instance, software is delivered to the client, its continued use is far from guaranteed. When the system had proven its due, Cammaert made sure it was properly documented, so that if he were to change to a different function, his successor could also benefit from it.

18.5.2   Gathering data

18.5.2.1   *Users participate*

That collecting data can be a tiresome task is no news. Particularly if the people who possess the data are not actively involved in the project, they may see no reason to waste effort getting the precise data asked for. During the group sessions with Fortuin, that problem was pre-empted by Callebaut's presence throughout the sessions. Whenever an employee would remark that the data needed to implement a new performance measure were hard to get, Callebaut would instantly order that they find out why those data were hard to get, and how they could be collected in the future.

18.5.2.2   *Using an intermediary*

The ways data were collected for Fleuren's simulations and to formulate Cammaert's model are interesting, so let us look at each in turn. To be able to perform useful simulations, Fleuren needed accurate estimates of machine breakdown frequencies, processing rates, etc. Some of the process engineers did not see the importance of such figures, and therefore could not be bothered to come up with precise estimates. Fortunately, one project team member, Van Gool, dedicated much effort to making Fleuren's life easier. Being a member of the team (which Fleuren was not), Van Gool was closer to the process engineers, and knew where to get the data needed. Smits said: 'Fleuren didn't have the connections to do that.' Van Gool was so enthusiastic about Fleuren's work that he became 'emotionally involved' in the simulations, and, where necessary, used that emotional involvement to influence the engineers into cooperating. By showing engineers how different the results of the simulation would be depending on whether they gave a 95 per cent or a 99 per cent machine reliability estimate, they began to see the importance of coming up with good estimates rather than seat-of-the-pants figures. Fleuren also addresses this issue in Chapter 12.

18.5.2.3   *Showing gaps in the data*

One of the difficulties Cammaert encountered was in obtaining figures for import duties and tax rates. For whatever reason, fiscal specialists did not provide him with precise figures, which he needed in order to complete the model; instead, they would say things like 'between 10 and 50 per cent'. However, by using the model and performing sensitivity analyses on the unknown parameters, Cammaert could establish which parameters really mattered and when each parameter became important. If it turned out that his product sourcing decisions should depend on whether a particular tax rate was below or above 35 per cent, he could go back to the fiscal department and ask them which was the case. Apparently, the fiscal staff felt more comfortable with this type of question, and could now give Cammaert more useful answers. Cammaert did remark that obtaining and interpreting the cost data and financial data was facilitated by the fact that his collaborator was a financial analyst.

18.5.3 Communication

18.5.3.1 Differentiated responses to OR, even within the same client organization

Different people, even within the same organization, can react very differently to OR. As the composition of the design team changed, Fleuren had to convince every newcomer of their contributions. During each of the subprojects, however, the people with whom Fleuren directly dealt did not change. What is interesting is the difference in reaction to Fleuren's simulation results. Process engineers on the project team often took Fleuren's results as the truth, eliminating the need for further discussion among themselves about 'how the plant would behave if . . .' . On the other hand, when the results were presented to people from outside the team, who had not been involved in the design of the plant, reactions were sometimes much more sceptical; people found it hard to accept that simulations could provide very useful insights if the input data were not exactly correct. Smits acknowledged that how people react to OR is very variable; although he is convinced of the benefits, he knows of many other people who, in his place, would not have seen the need to perform simulation studies. Callebaut adds that, in any project, it is important that the consultant and the client get on well together; establishing this is the other main purpose of a first meeting, besides discussing the project itself.

18.5.3.2 A project champion in the client organization

Recall the difference between Gray's positive attitude towards OR and the dismissive attitude displayed by his successor. Gray recognizes that many managers do not have a quantitative mindset; what surprises him is how many managers do have an analytical background but have 'deserted', too often as a result of 'intellectual laziness'. Brown recognizes that Gray and other clients often enjoy talking with consultants, and that this can be an important reason for them to initiate projects. Gray realizes, however, that his enthusiasm is very personal. Indeed, his successor was not interested in OR, and would not let Brown change his mind.

18.5.3.3 The client invests in understanding the consultant

A sceptical attitude towards OR may pose a challenge, but not an insurmountable one, which is clearly illustrated by Voûte. His opinion is unambiguous: 'I have an aversion to anything mathematical. Just say it in plain English, I always say.' Having no prior experience with OR, he found that it required an investment to understand the OR consultants; he believes this should not be the case, that OR consultants should improve their communication style.

> I think the big problem is that the OR consultant doesn't make it sufficiently clear what he's doing. He shouldn't use formulae and programmes, because any client who isn't an OR person himself is wary of anything that's like a black box.

An important consideration here is that Voûte and his staff have to understand the system and its philosophy well enough to be able to explain it to their own clients. Over the years, Boender has become sufficiently expert in the realm of pension fund management that he can compete with non-OR consultants traditionally more closely associated with pension funds.

18.5.3.4 *Complexity may frighten*

Capturing the interest of future users proved difficult in the RGD project. When selecting a new target group, Wassenaar did check whether the problem, in its new incarnation, was considered relevant; this was widely confirmed. But to get people to use the system will still take some time and effort, because, as Wassenaar explains, 'they haven't really been involved, it isn't their project. It's something which has been invented and of which they see it could be useful, but the complexity frightens them.' Some of the regional staff, the current target group for the system, were closely involved in its development, and they are enthusiastic. Others, however, were not involved in the 20 sessions leading up to the development of the system. It is turning out to be difficult to persuade them of its potential. ORTEC and Wassenaar intend to deal with this issue by organizing information sessions and by closely supervising a number of office allocation projects with people actually using the system.

18.5.3.5 *Starting with the user interface or with the 'inside' of a system?*

According to Boender, this problem was caused because they had started with the 'inside' of the system, the algorithmic part, leaving the design of the user interface until later. Usually, ORTEC takes the opposite approach. When the consultants are dealing directly with planners who are future users of a system, they start by building the input and output routines, to give the users an impression of what the system should be able to do, and only when the users' interest has been awakened, do they develop the mathematical inside of the system. Boender stated:

> I blame myself that we aimed too high, in the early part of the project we tried too hard to squeeze everything in, while one of ORTEC's strong points is to always start with something small and then expand it. Now both the RGD and

we believe that one of the reasons why the system isn't taking off as fast as we'd like is that it's too complex.

When asked why this was allowed to happen, Boender hesitantly responded 'I think it's because the initiative was theirs'.

18.5.3.6   Understanding the threat to the users

Dealing with future users of a planning system is always a sensitive affair, particularly when that system will replace some of the current planners. One of the reasons for Gray's satisfaction with ABC is the way it approaches planners. Although several of the systems ABC has introduced at Transco had the potential to be threatening for planners, by making a number of them redundant, the consultants managed to identify closely with the planners, and gain their full cooperation. Brown is very conscious of the challenges getting closer to planners sometimes poses. For instance, he has noticed a tendency by OR consultants to underestimate the complexity of some standard OR concepts for planners. A concept as sensitivity analysis, for example, is harder to convey to planners than is often realized.

18.5.3.7   Coaching future users

Other consultants are also familiar with these issues. In projects involving implementation of software to be used on a routine basis by planners, the process of training and coaching them is often considered, explicitly or implicitly, the responsibility of the OR consultants. Obtaining a planner's cooperation requires that management give him or her enough time to work with the consultants. Van Looveren recommends making that very clear from the start, to prevent problems later on. In the Oilco case, this went smoothly, but whenever a planner had doubts about the project, it was Bookman, the consultant, who had to coach the planner along, e.g. by being present during the first days after the actual changeover to the new system, and by installing a modem link with the BOR office. As Bookman summarized his coaching of the planners: 'I'll jump first, you just follow me'.

18.5.4   Expertise of client and consultant

18.5.4.1   Some clients are knowledgeable about OR

Clients of OR projects can have any degree of knowledge about OR. We have mentioned Cammaert, who was sufficiently knowledgeable about OR techniques to have formulated a linear programme himself. We have mentioned clients such as Smits, Gray, and Wassenaar, each of whom had a clear idea of the possibilities of OR, even if they were not familiar with the actual techniques used. Recall that it was Wassenaar who, by asking questions about the precise working of the algorithms, was able to suggest an approach more tailored to the problem at hand. And we have mentioned clients who knew little or nothing about OR, such as Voûte and Van Dedem. Clearly, the client knowing about OR is by no means a necessary condition for a project to be successful.

18.5.4.2 *OR projects can be perceived as risky*

Conversely, the consultant may or may not be an expert in the precise domain concerned. BOR consultants clearly are experts in the realm of cutting problems, which can have the important benefit of reducing the risk as perceived by the client. The support for the Oilco project generally came from people who were not high enough actually to authorize it; the people above agreed to the project but largely delegated it to the people below. At a higher level within the firm, a large-scale project such as this one entails a certain amount of risk. Bookman said:

> A client deciding to initiate a project like that is touching the organization of the firm, the way people plan, the way people do business, that man is taking a personal risk. It's easier to sell desk chairs for the same amount, there's no risk involved in that.

18.5.4.3 *The consultant has to learn about the client's unique problems*

Other consultants may be experienced in the general domain involved without being expert on the particular problem under consideration. Fleuren was a frequent user of simulation in the context of production processes, without being a specialist in the production of liquid crystal displays. Fleuren stated: 'If you don't include the material handling system, you can never advise your clients. You surely have to have some feeling yourself for which elements matter in a plant like that.' Fortuin, also with a wide experience of production processes and having studied performance indicators, had not actually implemented performance indicators in a plant before. And Boender, at the start of the joint venture with AMEV, had some knowledge of pension fund management, but was not yet an expert in the field. Finally, there were cases in which the consultants had no prior knowledge or experience with the particular problem concerned. Before the projects with the RGD, Boender had not worked on office space allocation-type problems, nor had Brown any prior experience with medium-term labour planning for maintenance personnel.

18.6 What did the client learn?

A very important but often underrated benefit of many OR studies is the increased understanding the client gains of his or her own business. Although the economic value of this learning is often impossible to measure, insight is frequently mentioned as the most important result of OR projects. Unfortunately, this learning effect is very poorly understood. It is not clear what types of insights clients gain, nor when and how these occur or can be made to occur. To illustrate the importance and the diversity of the phenomenon, we take a brief look here at what some of the clients learnt during their projects.

18.6.1 Monsanto

18.6.1.1 *Gathering systematic data allowed comparisons across products*

Cammaert obtained various types of useful insights from the project. To begin with, gathering the data already proved a valuable exercise.

> From gathering this cost data, it transpired that some product formulations cost about as much as what they sold for. We'd never seen things from that perspective before, because we were now forced to collect these data in a systematic way, so that they could be compared with each other, otherwise you cannot use them in a model.

Furthermore, it became clear that there were important differences in the relative importance of each of the types of costs involved. The ratio between transport costs, production costs, and duties and taxes appeared to be in the order of 1 to 10 to 100, which strongly suggests that duties and taxes should be the dominant factor in sourcing decisions.

18.6.1.2 *Making decision rules explicit and testing them*

An additional benefit of the modelling exercise was that it made explicit many of the rules Cammaert and his staff had been using in making sourcing decisions, and therefore allowed them to be critically checked.

> The results were rather impressive, particularly because the model showed us things we'd forgotten, we hadn't found those with our logical reasoning. That's not surprising, when you think about it, we work with so many daily constraints; we think that something isn't allowed, though we've forgotten why, but unless you explicitly include that in the model it won't take that rule into account. That way we found certain routes of which we thought 'that can't be possible', but analysis showed that we'd simply overlooked a few constraints. Then you wonder, 'if we'd forgotten them, were they really essential?' So basically that model has helped us to better understand our business.

Cammaert summarized by saying 'you have these rules from the past, that never get checked. The model allows you to check these rules.'

18.6.2 Compco

18.6.2.1 Making explicit trade-offs

Fleuren explained the goal of the simulations as getting insight into the behaviour of the production system, in terms of capacity, work in process and throughput times. The results of the simulations were often presented in the form of graphs of the trade-offs between these key variables. Smits knew that there would be capacity losses due to congestion effects, but the team had no idea exactly how high these losses would be. After the first simulation, the team knew that capacity losses would be about 11 per cent. That first simulation also resulted in a major insight for the project team: the perfectly balanced line it had initially designed would have a terribly poor performance, as every process step would turn out to be a bottleneck, not just the most expensive steps. Fleuren said:

> The interesting thing is that people in a design environment are inclined to balance the line perfectly, but what happens then? That line's behaviour is truly hopeless. . . . That's a way of thinking that grows over time. As consultant I've noticed that they have gradually learnt to think along those lines.

18.6.2.2 Finding design flaws

Various other insights occurred as a result of Fleuren's work. For instance, the data suggested there would be a relatively large number of 'short' machine interruptions and only a few longer ones. Therefore, it is important that the maintenance staff have a short response time. As a result of the simulations, the maintenance staff and a large collection of spare parts are now stationed on the shopfloor rather than in a separate workroom. Smits said: 'with hindsight, it sounds obvious'. Also, in the initial design, one particular material handling system had been specified. Fleuren, however, felt that the variability inherent in that system could be significant, and when he compared his simulation results under a 'fast' and a 'slow' scenario, the difference in capacity could amount to 30 per cent. This was a total surprise for the project team. As Smits said: 'The impact of the material handling system was shocking to me.'

18.6.3 RGD

18.6.3.1 Becoming aware of the problem's unsolvability

Perhaps the most important lesson in this project was that the office space allocation problem was simply unsolvable, when taking all the criteria that appeared relevant into account. Rather than find the best of a large collection of admissible allocations, the system would have to concentrate on finding the allocations violating the fewest criteria and letting the actual decision maker choose between them.

18.6.4 AMEV

18.6.4.1 Counterintuitive findings

Voûte and his team gained several fundamental insights into the business of managing pension funds. For instance, it was previously believed that investing in real estate was a reliable way for a pension fund to hedge against inflation. However, studies with the ALM system showed that, depending on the type of people insured by the fund, real estate could be a poor hedge, and that shares could be much more reliable. An insight that grew during the cooperation with Boender concerned the relative importance of the revenues from contributions and the revenues from investments: this turned out to be in the order of a 20 to 80 per cent ratio, suggesting that decisions concerning the investment policies of a pension fund could sometimes have a significantly larger impact than those concerning contribution schemes. More generally, Voûte feels much more confident and comfortable discussing such issues with his clients as a result of the project: 'You can communicate much more easily. And you can show things. You know much better what you're talking about.'

18.7 Conclusions

18.7.1 Summaries

18.7.1.1 An OR project as a 'kayak trip'

Perhaps the most important conclusion to be drawn here is that there is indeed much more to the practice of OR than meets the eye. There are many issues, of which we have only discussed a few here, which are critical to the success of any OR project, but that are hardly, if at all, mentioned in case descriptions. Performing a successful project involves much more than simply applying techniques, and projects are rarely the linear progression that case descriptions frequently suggest they are. To use Bookman's words: 'it's like a kayak trip. You can keep your head perfectly above water, but from time to time you need to take a bend. And when the water gets very calm, the waterfall's usually just round the corner.'

18.7.1.2 Practitioners have difficulties articulating their practice

We have seen that the practitioners with whom we spoke are all capable of performing highly successful OR projects, but they all found it difficult or impossible to articulate exactly what it was that made them successful. This state of affairs is obviously a major obstacle on the road to providing better education for students of OR, a better preparation for a career as an OR practitioner. What we need to do, as a profession, is capture the knowledge of experienced practitioners by documenting their work much more accurately than is commonly done, focusing especially on those issues they themselves do not write about. Incidentally, this situation is not unique to OR, but appears to be a general phenomenon in many professions. In this context, Argyris and Schön (1974) refer to the discrepancy between 'espoused theory' (what do practitioners say they do) and 'theory in use' (the theory that can be inferred by observing what they really do). Much of what practitioners do is in fact based on tacit knowledge, making it difficult for them to express themselves. To prepare students for practice, we need to teach them about the 'theories in use' of successful practitioners, but what we actually teach them are the 'espoused theories'.

Let us quickly look at how all this relates to three issues of concern to many within the OR community: the 'crisis debate' (the subject of Chapter 17), the increasing marginalization of OR departments in many business schools, and the fact that so many OR graduates abandon their field so soon after graduating.

The crisis in OR: Authors contributing to the crisis debate during the last five decades have done the profession a service by calling attention to the need to emphasize OR practice more heavily, resulting in several laudable initiatives. On the other hand, the resulting polarization between theory and practice has tended to overshadow the fact that what is really needed is a theory *of* practice. The literature is replete with statements that OR practitioners need to possess (more) social skills to reduce the frequent communication, political and other problems that occur in OR projects. But, if we are to teach students these skills, some formalized body of knowledge is needed about these skills, an 'epistemology of practice'.

Marginalization of OR departments: There is a trend among business schools to shift the emphasis in their teaching away from quantitative skills, resulting in downsizing and closing of many OR departments. Generally, MBA students are far more likely to end up as the client of an OR project than as the consultant. From the projects discussed here, we see that the client's knowing about the techniques of OR is not a necessary condition for a project to be successful (though it may sometimes help). At least as important is that the client has a good understanding of how to manage the process by which the project is executed, i.e. precisely those issues not normally described in OR cases. So, richer descriptions of cases are a necessary (but not sufficient) condition for training future practitioners, as well as for educating future clients!

Apprenticeship in OR practice: Lastly, we should emphasize that we definitely do not claim that OR practice is something that can be learnt from a book. Good descriptions and a good theory of practice may help a lot, but experience will always remain an essential learning mechanism. One approach often employed in management consulting firms is that of apprenticeship: by pairing a junior consultant with a senior consultant, the former can quickly learn a lot from the experience acquired over the years by the latter, and a lot of typical beginner's mistakes can be avoided. People who have to start practising OR as an individual without such an apprenticeship, such as the Lone Rangers, can easily, by making a few minor mistakes in their first projects, turn the tide in their organization against them, leading them and their organization to abandon OR much too quickly.

18.7.1.3   The need for serious research on the practice of OR

To conclude, we strongly believe that there is an urgent need for further, serious research on the practice of OR. One reason why such research is rare probably lies in the fact that the subject matter and the appropriate methodologies for such research are generally not those for which people within the OR community are well equipped to deal with, while scholars in other disciplines with the necessary skills generally have no particular incentive to apply those skills to studying OR. Although trying to bridge two gaps at once, that between theory and practice of OR and that between OR and other disciplines, is a challenging prospect, in our experience it has proved highly rewarding and stimulating, and we hope that many others will have similar experiences.

Acknowledgements

We would like to thank the OR consultants and their clients involved for their willingness to discuss these projects with us and their openness in doing so. We hope that this chapter reflects something of the enthusiasm with which they conduct and talk about their work.

Notes

[1] Among others, the creation of the journal *Interfaces*, devoted entirely to OR practice, the Edelman competition for excellence in OR/MS practice, and the recently established EURO award for the best applied paper.

[2] The names of some of the individuals and companies involved have been camouflaged; also, in some cases, certain numbers have been changed.

[3] Some of the more technical details are described in Chapter 12.

[4] Chapter 13 contains a more detailed discussion of the ALM system which resulted.

References

ARGYRIS, C. and SCHÖN, D. A. (1974) *Theory in Practice*, San Francisco: Jossey-Bass.

CORBETT, C. J. and VAN WASSENHOVE, L. N. (1993) The natural drift (what happened to operational research?), *Operations Research*, **41**, 625–40.

CORBETT, C. J., OVERMEER, W. J. A. M. and VAN WASSENHOVE, L. N. (1995) Strands of practice in OR (the practitioner's dilemma), *European Journal of Operational Research*, **87**, 484–99.

FORTUIN, L., VAN BEEK, P. and VAN WASSENHOVE, L. (1992) Operational research can do more for managers than they think!, *OR Insight*, **5**(1), 3–8.

CHAPTER NINETEEN

Epilogue

L. FORTUIN, P. VAN BEEK and L. N. VAN WASSENHOVE

19.1 The proof of OR is in the eating

Learning about OR practice is best achieved by actually doing it. The next best thing is to read about it in cases – provided these cases are a true reflection of the richness and complexity of what a seasoned practitioner really does. Of course this goal can never be achieved. The diversity and complexity of many OR interventions can never be captured in a few pages. No problem. If OR students and novices were to be confronted with a well-selected set of rich cases by instructors with practical experience, they would certainly be willing and fully able to appreciate the challenges of the process of an OR study. When combined with one or two closely supervised real-life projects, this would certainly prepare them better to jump into the water and become successful practitioners.

The future of OR is not past (Ackoff, 1979). The future of OR is not necessarily bright either (Rinnooy Kan, 1989). But the future of OR as a discipline is certainly ours. It will be what we choose to make of it. In the views of the *troika*, as we, the editors, are still called occasionally, OR has to take its subject matter, i.e. applications, much more seriously. The process of OR is poorly documented and poorly understood. There is a huge need for much richer case descriptions. There is also a huge need to apply knowledge engineering to our successful practitioners in order to understand better what it is they really do, as opposed to what they say or write about. We hope this book offers a modest contribution to better understanding of OR practice. If nothing else, we sincerely hope we have been able to convince the reader that successful practice, and training future practitioners, will be key to the future of OR as a discipline. Successful practice therefore deserves much more attention, from our professional societies as well as from universities and training institutes.[1] What constitutes successful OR practice should be the subject of serious research efforts.

We truly believe in the high success potential of OR as a discipline provided that OR takes its *subject matter* seriously – we have no doubts about the *technical part* of the discipline being taken seriously. Taking its subject matter seriously also means acknowledging the rich variety of potential applications and the many ways in which OR practitioners can help decision makers structure, understand and solve their problems and messes.

19.2   Process focus or tool orientation?

Just like there is a need for more tolerance, appreciation and exchange between theoretical developments and practice – and for 'management engineers' to bridge the potential gap as discussed in Chapter 17 – there is also a need for better understanding of the different clusters of OR applications, as well as the different skills required to operate in them.[2] There is no such thing as a good application or a bad application of OR! More technical OR studies are not superior to softer OR interventions or vice versa. We sincerely hope that the parts of OR applications presented in this book illustrate the richness and variety of our discipline as well as its health and growth potential. We hope the parts also illustrate the need for different types of OR practitioners (and therefore more respect and tolerance between them). To re-emphasize this point let us consider a simple 2×2 matrix (Figure 19.1):

- on the horizontal axis we indicate the OR practitioner's required knowledge of the application domain (e.g. chip design or plant location problems);

- on the vertical axis we indicate the necessity for the client to understand OR tools and/or approaches.

We can then position our five parts of OR applications in the matrix and show increasing needs for closer attention to process or to tools. For example:

- *Clean-room OR* clearly requires good technical and computer science skills from the OR practitioner whereas the actual client may have no knowledge about OR whatsoever. There is no need for the OR worker to have a deep knowledge about the domain (e.g. chip design) as long as client and OR worker can jointly define some intrinsic optimization problems (e.g. Hamiltonian path problems). Of course, this view is black and white. Knowledge of the domain by the OR worker and knowledge of OR by the client obviously won't hurt! But it should be clear that the OR practitioner does not require excellent process skills to be successful.

- *Commodity OR*, on the other hand, requires a knowledgeable and sophisticated user with good technical skills (e.g. in LP) as well as good

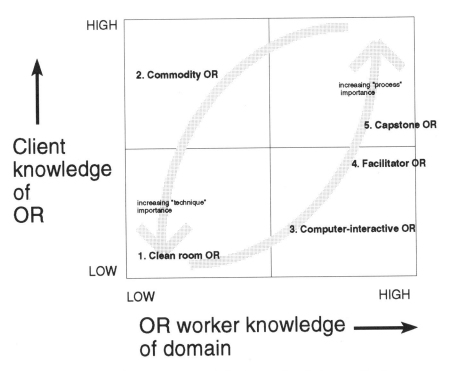

Figure 19.1 OR application parts and the interaction between client knowledge of OR and OR worker knowledge of domain.

computer skills (spreadsheets, graphics packages, word processors). The ease of use of current tools (e.g. LP in spreadsheets) contains the risk of misuse. There is a need here for the OR worker to train the user and develop foolproof tools. Domain knowledge is not absolutely required.

- *Computer-interactive OR* requires reasonably good technical skills from the OR worker as well as good computer skills (user-friendly interface design, graphics and animation). In addition, the OR worker requires good domain knowledge in order to be effective in providing 'knowledge engineering' for the user about the intricacies of the specific application. Indeed, problems here often arise because of insufficient attention to specific customer knowledge and needs.

- *Facilitator OR* requires a user with at least elementary notions about OR techniques and approaches. The OR worker requires reasonable technical skills but, above all, a broad (domain) knowledge and experience base and excellent interpersonal (soft) skills. Surprisingly, many applications fail not because of poor soft skills (practitioners lacking soft skills don't even get a chance!) but because of too much emphasis on the soft issues at the expense of up-to-date knowledge and use of OR tools. Even soft OR consultants should know and apply the powerful modern tools of our discipline!

- *Capstone OR* requires fair technical skills from both the OR worker and the client. The OR worker also needs excellent interpersonal skills (political flair) and a broad knowledge of several domains. Many failures in this complex area are due to insufficient domain knowledge (underestimating some parts) and lack of feeling for what is acceptable (*Realpolitik*).

The above is just an illustration of the usefulness of trying to understand and cultivate the richness of our discipline. It shows how different applications put different requirements on clients' and OR practitioners' knowledge as well as on how they need to interact and cooperate. From this, conclusions can be drawn about how to educate and train OR practitioners and users. It also suggests that the question of whether OR curricula should be housed in a mathematics and statistics department, an engineering department or the business school is superfluous. OR can fit everywhere and we should be proud of that. And there is certainly room for tolerance and cross-fertilization between different subfields and between more technical versus softer approaches. Although we feel it is absolutely essential to develop further this type of exercise, it is outside the scope of this book.

19.3 PR for OR

Fierce competition forces organizations towards more effective decision making. OR definitely adds value here. Unfortunately, many managers are unaware of OR and many others think of it as mathematical wizardry inside mysterious black boxes. Obviously, decision makers will only foster OR approaches when they are aware of them and believe in them. It follows that the OR profession faces the extremely important task of making successful OR applications much more visible. Fortunately, the professional OR societies seem to have realized this and are (finally) beginning to develop serious initiatives in this direction. We hope this book is also a modest contribution.

Shrinking budgets play havoc with OR curricula. OR departments are disbanded and OR courses scrapped from many programmes. This trend needs to be reversed (cf. Ranyard *et al.*, 1995). Here again one needs to seduce the potential client by showing the added value of the discipline. Rich descriptions of successful applications, i.e. cases, are a good way of doing this. As such, the current book can help show OR students that their discipline is not a branch of mathematics. Equally important, it can show the value of OR to non-OR students in other programmes. Indeed, they are the future managers and potential users of our discipline.

OR definitely adds value. This book testifies to this. It does so by its models and techniques but, above all, by its way of thinking. However, it is not always easy to convince others. For instance, in our clean-room OR part, how does one convince a plant manager that flexible customer response depends to a large extent upon a clever, state-of-the-art, combinatorial optimization algorithm embedded somewhere deep down in a computer-integrated manufacturing system? And how does one make a bunch of functional managers realize that their successful cross-functional cooperation is the result of a learning process in which an OR facilitator used the OR way of thinking to lead them through a set of scenarios showing the tremendous benefits to all from closer cooperation? Difficult, but absolutely necessary! We need to think of better ways to market the successes of our discipline and its value added, both to managers and to students in non-OR programmes.

Although we feel that OR has many faces and multitudes of potentially successful areas of application and growth, we are also painfully aware of its boundaries. As Socrates said: 'Real wisdom is about knowing how much there is one does not know.' Similarly, wise OR practice requires understanding of OR's limitations. That is, whenever appropriate, get other disciplines on board[3] or tell the client honestly that OR is not what he or she needs in this particular case. You will be doing your discipline a favour. OR also requires modesty. It is just one of the many (potentially useful) ways of viewing a (decision) situation. Sometimes the OR view is the best, many times it is not. Often several complementary views from different disciplines are most adequate. Similarly, many decision makers still use seat-of-the-pants (intuitive) approaches and get away with luck. Some use a more scientific (structured) approach but only part of those are comfortable with models and numbers! A well-known OR professor recently expressed the above considerations by paraphrasing an old French saying:[4] *Tout ce qui brille n'est pas de l'OR.*

19.4   Old habits . . .

After almost 10 years of efforts at promoting the successful practice of OR, the editors of this book set out to produce a *pièce de résistance* before disbanding as a team. The casebook they would produce would be different in many ways. It would show the many faces of OR, i.e. the large variety and richness of the discipline by selecting non-traditional applications. The case descriptions would also have to be different from the traditional 'application of a technique to a given problem' or 'application of a linear sequence of steps from problem definition to solution implementation'. These views are too limited and potentially damaging to the image of the field. The *troika* would therefore urge the contributing authors to provide them with much richer descriptions about what it was they really did, i.e. to describe the process much more than the techniques. Finally, the *troika* would also provide the reader with some 'behind the scenes' information by adding a chapter based on interviews with practitioners and their clients to illustrate the true richness of an OR intervention as well as all the potential pitfalls.

Only the reader can be the final judge of whether the editors have succeeded in making this book different. As stated earlier, we all have our biases and old habits die hard . . .

Notes

[1] The editors are aware of many excellent efforts by societies and individuals in different countries. However, for every successful effort they also see at least half a dozen cases where an awful lot remains to be done. See also Ranyard *et al.* (1995).

[2] The diversity of the cases in this book (and elsewhere) shows that many different types of OR exist, each with its own requirements on OR practitioners' skills, and its own inherent characteristics and dangers. To get a better grip of this diversity we used parts to structure the book. Our clustering is meant to be thought stimulating rather than definitive or limitative as a classification! However, our clustering does suggest that serious research is required to understand better successful OR practice.

[3] Remember OR's origins before and during World War II as a truly cross-functional discipline!

[4] 'Not all that shines is OR (gold)', D. de Werra during his EURO Gold Medal acceptance speech, Jerusalem, July 1995.

References

ACKOFF, R. L. (1979) The future of OR is past, *Journal of the Operational Research Society*, **30**, 93–104.

RANYARD, J., CRYMBLE, W. and FILDES, R. (1995) Study finds OR in good heart (Whilst thinking it might do better), *OR Newsletter*, 8-16 (July).

RINNOOY KAN, A. H. G. (1989) The future of operations research is bright, *European Journal of Operational Research*, **38**, 282–5.

Who's who in this book

Jacques Benders holds master's degrees in mathematics and physics, and a PhD in mathematics. After 10 years of experience in the industrial application of statistics and operational research, he became a full-time professor of mathematics at the Eindhoven University of Technology, in charge of teaching and development of mathematical and computational aspects of operational research methods. Now retired, he is still active as a consultant, stimulating the sound application of these methods where possible in industrial practice.

Guus Boender is professor of OR, in particular decision support systems, at the Erasmus University Rotterdam, and partner of ORTEC Consultants bv, a company with 70 consultants on quantitative decision support. His theoretical research areas and publications mainly concern stochastic modelling and stochastic optimization. These subjects serve well his current interest of applied financial modelling, in particular asset liability and risk management. In this field he is active as an applied operational researcher for a large and growing number of pension funds, insurance companies and banks.

Frits Claassen works as a research scientist in the Department of Mathematics at the Agricultural University Wageningen. After his graduation from the same university, he worked there for two years as a temporary researcher on production planning and scheduling, in a project supported by the government. In his work he is primarily involved in the application of OR in the agricultural and environmental sciences.

Charles Corbett is currently a PhD candidate in production and operations management at INSEAD; for the academic year 1995–6, he holds a Visiting Scholar position at the Owen Graduate School of Management at Vanderbilt

University. He is engaged in research into the practice and process of OR, in close cooperation with OR practitioners. His other main research area is supply chain coordination, in which he is performing theoretical work and consulting on interfirm supply chain improvement projects.

Joachim Daduna has taught business administration and logistics at the Fachhochschule Konstanz since 1994. Before that he worked for more than 10 years in public transport, first in software engineering for the Hamburger Hochbahn Aktiengesellschaft and later as a consultant for Dornier GmbH, a subsidiary of the Daimler-Benz Group. He studied business administration, economics and social sciences and holds a master's degree and a PhD in business administration. The main fields of his research and also of current consulting activities are public transport and logistics, particularly referring to OR applications in practice.

Rommert Dekker is a full-time professor of OR at the Erasmus University Rotterdam. He holds an MSc (1981) and a PhD (1985) in mathematics from the State University of Leiden and a degree in industrial engineering from the Technical University of Twente (1989). From 1985 till 1991 he worked with Shell Research and Shell International on maintenance optimization, reliability and logistics, and was responsible for the development of several decision support systems. His present interests also include remanufacturing logistics, spare parts management, distribution optimization and container logistics.

Dick den Hertog is a consultant for OR at CQM in Eindhoven. At the moment he is also secretary of the Dutch OR Society. In recent years he has mostly worked in the field of physical distribution, doing both strategic and operational projects, and in the field of production design. He is especially interested in product optimization with CAD tools. His PhD thesis on interior point methods (Delft University of Technology, 1992) was published by Kluwer in 1993. Since then his interest has moved to applied work.

Matthijs Dijkstra was one of the initiators and stimulators of the project for aircraft maintenance personnel that has been carried out for KLM-VOC. In 1990 he left academia. He is now employed as an information manager with Bouwcenter Group.

Hein Fleuren is manager of the OR group at CQM. His main interest is the practical application and implementation of OR in a broad sense. Until 1995 he was a consultant at CQM, collecting seven years of experience in consultancy projects. His main fields of experience are vehicle routing and discrete event simulation of production systems. On the subject of vehicle routing he wrote a dissertation at the University of Twente in 1988. Furthermore he holds a master's degree in applied mathematics and mechanical engineering.

Leonard Fortuin works part time at the Eindhoven University of Technology. Logistics, especially of after-sales service, and performance management are his main fields of interest. Until May 1993 he was a consultant for OR with the Centre for Quantitative Methods in Eindhoven. He holds a degree in electrotechnical engineering and a PhD in technical sciences. Promotion of OR in practice, not only by application of OR for the solution of industrial problems but also by publications in professional journals and newspapers, is one of his current activities. His experience over the past 20 years has been gained primarily in consultancy projects throughout a multinational electronics company.

Dik Habbema studied applied mathematics at the Technical University of Eindhoven. He joined the Statistical Consultation Department at the Medical School and University Hospital of Leiden University. In 1975 he moved to the Department of Public Health at the Erasmus University Rotterdam. He developed research programmes in theory and applications of clinical decision analysis and in the evaluation of health services and disease control programmes, nowadays also called medical technology assessment. Research projects include cost-effectiveness studies of cervical cancer and of breast cancer screening, technology assessment of liver transplantation, evaluation of river-blindness control measures, and clinical decision analysis studies of severe head injury, jaundice diagnosis, fertility-enhancing surgery, epilepsy, lumbar herniated disc and intracranial aneurysms. His current positions are director of the Center of Clinical Decision Sciences and of the Centre for Decision Sciences in Tropical Disease Control, and coordinator of the decision-making research programme at the Department of Public Health. He holds the Chair of Medical Decision Sciences at the Faculty of Medicine of the Erasmus University.

Eligius Hendrix works in the Department of Mathematics at the Agricultural University of Wageningen, in the section for OR. He is mostly involved in the application of OR in agricultural and environmental sciences. Specific interests are the application of global optimization methods and the modelling of uncertainty in decision models. He studied econometrics at Tilburg University and development planning at Erasmus University Rotterdam. After some work for the UNO and for Tilburg University, he joined the Agricultural University of Wageningen in 1987.

Leen Hordijk has a background in econometrics, graduated from Erasmus University Rotterdam, and received a PhD from the Vrije Universiteit in Amsterdam. During the last 15 years, he has specialized in large-scale environmental modelling, in particular in acid deposition in Europe and Asia. In Wageningen he holds a chair in environmental systems analysis and is director of one of the university's research institutes. He is an alumnus of the International Institute for Applied Systems Analysis (IIASA) in Austria.

Cor Hurkens received his master's degree in mathematics (with Prof. Benders) in 1984 and his PhD in 1989 on research in combinatorial optimization. Currently he is an assistant professor in combinatorial optimization at the Eindhoven University of Technology. His research interests are in combinatorial problems with real-life applications such as scheduling, cutting stock and routing problems, particularly the worst-case analysis of heuristics for these problems.

Karl Inderfurth (born in 1948) studied economics in Bonn. He moved to the Free University of Berlin where he graduated to Dr. rer. pol. in 1975, and completed his Habilitation in Business Administration in 1981. For a couple of years he served as a top management assistant at the headquarters of a multinational pharmaceutical company and as the head of a global logistics project. In 1988 he returned to academia, to become professor of OR at the University of Bochum. Later he moved to Bielefeld University, and since 1994 has held a chair for production and logistics at the University of Magdeburg. His research activities are directed towards analysis and management of risk phenomena in production and logistics. Inderfurth chairs the 'inventory' working group of the German Society of Operations Research (DGOR). He is co-editor of journals in OR and production economics.

Leo Kroon teaches OR and operations management at the Rotterdam School of Management. His main research interests include the application of quantitative methods and decision support systems to real-life problems.

Cindy Kuijpers received her MSc degree in applied mathematics at the Eindhoven University of Technology in 1993. Her thesis, based on a project carried out at the Philips Research Laboratories, was awarded the 1994 VVS prize of the Netherlands Society for Statistics and OR. Currently she is a PhD student at the University of the Basque Country, where she is involved in a research project on the application of genetic algorithms to combinatorial optimization problems related to Bayesian networks.

Richard Lukkassen is a consultant for OR at CQM in Eindhoven. Since 1990 he has applied OR in a wide variety of consultancy projects. Many of these projects were in the field of combinatorial optimization. He is especially interested in applying and implementing OR solutions in a practical environment. He holds a master's degree in applied mathematics.

Dirk Meier-Barthold (born in 1965) studied economics at the University of Bielefeld. After his graduation in 1991, he worked at the Faculty of Economics in Bielefeld in an 'information management' project. This project included an education programme for practitioners working in the middle management of well-known companies. Since July 1994 he has been an assistant to Prof. Dr Karl Inderfurth at the University of Magdeburg. His

PhD project is directed towards the analysis of the flexibility problem, especially for general planning methods and specific decision rules in material coordination. Furthermore, he is engaged in projects on applying OR methods to solve practical problems in production and logistics management.

Hans Melissen received his master's degree in mathematics in 1982 at Utrecht University. Since 1986 he has been employed by Philips Electronics, currently as a research scientist at Philips Research Eindhoven. His research interests include non-linear optimization, computational geometry (packing and covering problems) and the analysis and numerical treatment of Maxwell's equations. One of his other activities is as book review editor of *ITW Nieuws*, the newsletter of the Dutch Society for Industrial and Applied Mathematics.

Wim Overmeer is assistant professor of strategy and management at Stern School of Business (New York University) and at INSEAD (France). He holds a PhD in business policy and general management from MIT (1989), and master's degrees from Erasmus University Rotterdam (1978) and Delft University of Technology (1976). He also is a consultant with large and small consultancy bureaux in Europe and the United States. His research focuses on problems that managers face when designing and implementing a business strategy, more in particular on the best type of 'enquiry' and 'intervention' to attack such problems effectively.

Eric Poot studied horticulture at Wageningen Agricultural University. Since 1991 he has applied OR in several planning support systems, all in the field of production and distribution of horticultural products. Since 1993 he has worked as a product manager with a flower bulb company.

Jaap Roosma studied food science at the Agricultural University of Wageningen. He obtained his master's degree in 1992 on the development of a model for production planning in the fodder industry. After that he joined one of the largest fodder companies in The Netherlands, and continued to work on production planning. In 1993 he worked part time for the Department of Mathematics in Wageningen and for a consultancy firm in Gouda. As a full-time employee of that consultancy firm, he delivered in January 1994 a DSS for production planning in the fodder industry to that firm.

Marc Salomon is associate professor in OR at the Rotterdam School of Management, and part-time consultant with ORTEC. He teaches management science and logistics. His research interests are in production planning, inventory control and scheduling. At ORTEC he is involved in the development and implementation of decision support systems for production planning, inventory control and scheduling.

George Schepens is a half-time professor at the management department of the Faculté des Sciences Economiques et Sociales of the Facultés Universitaires ND de la Paix in Namur, Belgium. His main fields of interest are operations management, logistics and OR. He holds two degrees in engineering (Leuven and Gent), studied management at the University of Gent and obtained a CRB fellowship which financed graduate studies at the Sloan School of the Massachusetts Institute of Technology. He is chairman and partner at Beyers Innovative Software NV, where he is involved in the development of optimization systems and in the implementation of integrated production planning control systems.

Luc Schepens is an engineer in electromechanics at the University of Brussels (VUB). He also holds a degree in industrial management from the Catholic University of Leuven (KUL). Since 1987 he has worked as a consultant and project manager for Beyers Innovative Software in Antwerp, gaining a wide experience in industry planning problems in an international context. Most recent implementations under his project management are in chemicals, food and agricultural businesses. The different planning problems solved by Beyers Innovative Software range from strategic/tactical planning, through master production scheduling to detailed scheduling. Luc Schepens also participated in the European Union Esprit project no. 6588 'ICEP' (Integrated and Concurrent Enterprise Planning).

Ewout Steyerberg studied health sciences at the University of Leiden, specializing in medical statistics. Since 1991, he has worked at the Center for Clinical Decision Sciences of the Department of Public Health, Erasmus University Rotterdam. He is a consultant for medical decision analysis and general epidemiological and statistical methodology for the University Hospital 'Dijkzigt' Rotterdam. In this function he has been involved in the design and analysis of several clinical trials and other clinical studies. His special interest is in prognostic models in general and in logistic regression in particular.

Paul van Beek is professor of OR at the Agricultural University of Wageningen. Until 1982 he worked with Philips in Eindhoven, ultimately as manager of a corporate OR group. He holds a PhD in probability theory from the University of Bonn, Germany. He has published many articles on OR applications and is co-author of a book on optimization techniques. In the period 1986–90 he was president of the Dutch Society for OR. Paul van Beek is a member of the editorial board of *EJOR* and editor of *OR Spektrum*, the official journal of the German Society for OR. Currently he is chairman of the task force for logistics of the Netherlands Council for Agricultural Research.

Yolanda van der Graaf has been working since 1987 as a clinical epidemiologist at the Academic Hospital in Utrecht (AZU). Her main task is the design and supervision of clinical studies, in particular relating to cardiovascular topics (effectiveness of vascular interventions, diagnosis of vascular disease, risk of prosthetic heart valve failure in Björk–Shiley heart valve carriers). Part of her job is the education of medical students and methodological consultancy in various clinical studies. After her qualification as a medical doctor she worked for two years as a community health officer in the Republic of Seychelles (1980–2). From 1982 to 1987 she worked as a research fellow on the effectiveness of cervical cancer screening in The Netherlands. She has written a thesis and many international publications on this subject.

Lex A. van Herwerden is a cardiopulmonary surgeon at the Thoraxcenter Rotterdam. His main interests are heart valve surgery, echocardiography and interventional therapy for coronary artery disease. He has consulted experts in OR/decision making when confronted with the problem of defective heart valves. He considers the use of human tissue valves for heart valve replacement a fertile area for future cooperation of clinicians and OR consultants.

Anita Van Looveren holds a degree in econometrics (UFSIA) and followed the MBA programme at the Catholic University of Leuven. She worked for several years at the Catholic University of Leuven as a research and teaching assistant. In 1985 she joined Beyers Innovative Software, a software and consultancy company specializing in the development and implementation of planning systems using OR and artificial intelligence techniques. In 1989 she became a partner and head of the operations management division. Since 1994 she has been the managing director of Beyers Innovative Software. She has been involved in a wide variety of projects in the field of strategic, tactical and operational planning. Current interests are in the area of integrated solutions for master scheduling and detailed scheduling.

Jan van der Meulen has worked since 1993 as a clinical epidemiologist in the Department of Clinical Epidemiology and Biostatistics at the Academic Medical Centre – University of Amsterdam. His main fields of interest are clinical epidemiology and decision analysis. At present he participates in studies related to the effectiveness and costs of various diagnostic and therapeutic procedures and the development of guidelines for clinical practice. Other research areas include the early causes of cardiovascular disease. He studied medicine at the University of Groningen. In 1989 he completed a PhD thesis at the University of Utrecht on the functional organization of the motor system in children. After that he joined the Centre for Decision Sciences of the Erasmus University in Rotterdam, where he worked for three years as a research fellow on projects in the field of clinical epidemiology and decision analysis.

Jo van Nunen is chairman of the Decision Sciences Department of the Rotterdam School of Management and former dean of the MBI programme at Erasmus University. He teaches and supervises research in OR, decision support systems, logistics and telematics.

Cyp van Rijn recently retired from Shell Research where he held various management positions in process systems engineering. His research interests lie with advanced control, process optimization and logistics, and he was active in organizing many conferences in this area (IFAC, CACHE). He has a special interest in process reliability, maintenance and safety, and was founder of the European Safety, Reliability & Data Association (ESReDA) which he still serves as honorary president. He now works as an independent consultant for the European Commission, industrial companies and in education.

Luk Van Wassenhove is professor of operations management and OR at INSEAD, Fontainebleau. Before joining INSEAD in 1990, he held faculty positions at the engineering school of the Catholic University Leuven (Belgium) and at the Econometric Institute of Erasmus University Rotterdam (The Netherlands). His research interests are in modelling complex operational, tactical and strategic problems in manufacturing, distribution and services. Besides his theoretical work he is also actively involved in case development, executive education and consulting activities.

Manfred Völker is a branch manager in mathematical systems at the HanseCom Gesellschaft für angewandte Informatik und Informationsverarbeitung mbH, a share company of the Hamburger Hochbahn AG (HHA). In 1981 he finished his study with a master's degree in electrotechnical engineering. From then until 1987 he worked as a software engineer and project manager. Since 1987 he has been employed by the HHA. On taking over responsibility for the branch of mathematical systems at HanseCom in 1991, he began working in the field of vehicle and crew scheduling in public transport companies.

W. Henk Zijm (born 1952) is a full professor of production and operations management. He holds a PhD from the Eindhoven University of Technology and an MSc degree from the University of Amsterdam. Previously he worked as an assistant professor at the University of Amsterdam, as a full professor (part time) at the Eindhoven University of Technology and as a senior consultant in production and logistics management at Philips Electronics in Eindhoven. His research interests include manufacturing systems design, capacity planning and lead time management, shopfloor control and process planning, design and control of warehousing and materials handling systems. He has published more than 60 papers in international scientific journals and is a member of TIMS, POMS, EurOMA and ISIR.

Index